D1413936

GEOGRAPHY BASICS

GEOGRAPHY BASICS

Volume 2

Glossary
Appendices

Edited by
Ray Sumner
Long Beach City College

SALEM PRESS

PASADENA, CALIFORNIA   HACKENSACK, NEW JERSEY

The essays, glossary definitions, and appendices in this publication
first appeared in *World Geography* (2001), copyrighted by Salem Press.
New material has been added.

∞ The paper used in these volumes conforms to the American National Standard for Permanence of Paper for Printed Library Materials, Z39.48-1992 (R1997).

Library of Congress Cataloging-in-Publication Data
Geography basics / editor, Ray Sumner.
    p. cm. — (Magill's choice)
  Includes bibliographical references and index.
  ISBN 1-58765-177-7 (set: alk. paper) — ISBN 1-58765-178-5 (vol.
1: alk. paper) — ISBN 1-58765-179-3 (vol. 2: alk. paper)
  1. Geography. I. Sumner, Ray. II. Series.
G116 .G475 2004
910—dc22

                                                     2003018130

First Printing

PRINTED IN THE UNITED STATES OF AMERICA

CONTENTS

GLOSSARY

This glossary defines terms and concepts mentioned in the main text. Words printed in SMALL CAPITAL LETTERS *have entries of their own here.*

Aa. Hawaiian term (pronounced "ah-ah") that has been adopted for LAVA flows with rough, clinkery surfaces.

Geologist examining a large solidified flow of aa lava from Mauna Ulu on the island of Hawaii. (U.S. Geological Survey)

Ablation. Loss of ice volume or mass by a GLACIER. Ablation includes melting of ice, SUBLIMATION, DEFLATION (removal by WIND), EVAPORATION, and CALVING. Ablation occurs in the lower portions of glaciers.

Ablation, zone of. In a GLACIER, at the SNOUT of the glacier where ice is lost through melting and SUBLIMATION.

Abrasion. Wearing away of ROCKS in STREAMS by grinding, especially when rocks and SEDIMENT are carried along by stream water. The STREAMBED and VALLEY are carved out and eroded, and the rocks become rounded and smoothed by abrasion.

Absolute age. Numerical timing (in years or millions of years) of a geologic event, as contrasted with relative (stratigraphic) timing; a date that gives an actual age, though it may be approximate, of an artifact.

Absolute humidity. Mass of water vapor contained in a unit volume of moist AIR. Absolute humidity is usually measured as grams of water vapor per cubic meter of air. More important to geographers is the RELATIVE HUMIDITY.

Absolute location. Position of any PLACE on the earth's surface. The absolute location can be given precisely in terms of DEGREES, MINUTES, and SECONDS of LATITUDE (0 to 90 degrees north or south) and of LONGITUDE (0 to 180 degrees east or west). The EQUATOR is 0 degrees latitude; the PRIME MERIDIAN, which runs through Greenwich in England, is 0 degrees longitude.

Abyss. Deepest part of the OCEAN. Modern TECHNOLOGY—especially sonar—has enabled accurate mapping of the ocean floors, showing that there are MOUNTAIN CHAINS, or RIDGES, in all the oceans, as well as deep CANYONS or TRENCHES closer to the edges of the oceans.

Abyssal. Referring to the deep-OCEAN floor. The part of the CONTINENT under shallow water is the CONTINENTAL SHELF. At the outer edge of the continental shelf is a steep fall or CLIFF, called the CONTINENTAL SLOPE. Beyond this is the ABYSSAL PLAIN.

Abyssal plain. Broad flat areas beneath the OCEAN at the base of the continental rise, at depths ranging from 1,200 to 3,500 feet (2,000-6,000 meters). Individual MOUNTAIN peaks can occur on abyssal plains. Close to 40 percent of the oceans are abyssal plain.

Abyssal seafloor. Abyssal plains of the OCEANS lie beyond the CONTINENTAL MARGINS at depths greater than 6,000 feet (2,000 meters). They are thought to be the flattest areas on the earth and are carpeted with thick layers of SEDIMENT. Their greatest economic value lies in the metallic MINERALS that form part of these sediments.

Acclimatization. Gradual adjustment of living organisms, especially humans, to climatic conditions other than those to which they are accustomed.

Acculturation. Modification of a CULTURE when its people come into contact with another culture. Generally, acculturation occurs when INDIGENOUS PEOPLES come into contact with a technically superior culture, as during the period of colonial empires. The globalization of entertainment (movies and television) has acculturated many societies to American culture. Compare with TRANSCULTURATION.

Accumulation, zone of. In a GLACIER, the part where ice accumulates as new SNOW falls each year.

Acid deposition. *See* ACID RAIN.

Acid rain. PRECIPITATION containing high levels of nitric or sulfuric acid; a major environmental problem in parts of North America, Europe, and Asia. Natural precipitation is slightly acidic (about 5.6 on the pH SCALE), because CARBON DIOXIDE—which occurs naturally in the ATMOSPHERE—is dissolved to form a weak carbonic acid. In areas where heavy industry is located, oxides of sulphur and/or nitrogen combine with atmospheric moisture to produce sulfuric acid or nitric acid, respectively. In the worst-affected parts of North America and Europe, pH as low as 2.0 has been recorded, which is more acidic than lemon juice or vinegar. As a result, thousands of LAKES and STREAMS in North

America and Europe can no longer support fish; FORESTS in Switzerland, Germany, and Poland have been damaged extensively. Buildings also are eroded by acid deposition, so that cathedrals and monuments are being destroyed. Often called acid rain; however, SNOW, SLEET, and hail can also be acid.

Acoustic echo sounding. Also known as sonar, method of determining the depth of the OCEAN floor that measures the time of a reflected sound wave and relates that to distance.

Adiabatic. Change of TEMPERATURE within the ATMOSPHERE that is caused by compression or expansion without addition or loss of heat.

Advection. Horizontal movement of AIR from one PLACE to another in the ATMOSPHERE, associated with WINDS. See also CONVECTION.

Advection fog. FOG that forms when a moist AIR mass moves over a colder surface. Commonly, warm moist air moves over a cool OCEAN CURRENT, so the air cools to SATURATION POINT and fog forms. This phenomenon, known as sea fog, occurs along subtropical west COASTS. Advection fogs are common in San Francisco, especially in summer.

Aeolian. See EOLIAN.

Aerate. To supply with or expose to a gas.

Aeration, zone of. Area directly below the ground surface that contains some water as SOIL MOISTURE, but much of the pore space is filled with AIR in the spaces between the soil particles. At the bottom of the zone of aeration is the WATER TABLE.

Aerosol. Substances held in SUSPENSION in the ATMOSPHERE, as solid particles or liquid droplets.

Aftershock. EARTHQUAKE that follows a larger earthquake and originates at or near the focus of the latter; many aftershocks may follow a major earthquake, decreasing in frequency and magnitude with time.

Agglomerate. Type of ROCK composed of volcanic fragments, usually of different sizes and rough or angular.

Agglomeration effect. Certain industries can obtain cost advantages by locating production among functionally related industries or activities. This occurs chiefly because transport costs are reduced. During the INDUSTRIAL REVOLUTION, manufacturing became concentrated near coalfields for this reason.

Aggradation. Accumulation of SEDIMENT in a STREAMBED. Aggradation often results from reduced flow in the channel during dry periods. It also occurs when the STREAM's load (BEDLOAD and SUSPENDED LOAD) is greater than the stream capacity. A BRAIDED STREAM pattern often results.

Agribusiness. Modern type of commercial agricultural production in which a company owns large areas of farmland and is concerned with not only the production of agricultural commodities, but also their transport, storage, processing, and distribution. The word is a combination of "agriculture" and "business." In the United States, agribusiness accounts for about one-fifth of the GROSS DOMESTIC PRODUCT.

Agricultural Revolution. Also known as the Agrarian Revolution, historical change from a nomadic lifestyle of HUNTING AND GATHERING or nomadic herding to a sedentary one based on the growing of crops. Scholars believe that this change first occurred in the area of the Middle East known as Mesopotamia at least eleven thousand years ago. Grain, varieties of either wheat or barley, was harvested for human consumption and for feeding domesticated animals, such as cattle or sheep. On other CONTINENTS, agriculture was practiced with different crops: squash and corn in the Valley of Mexico, LEGUMES in Southeast Asia. These changes made possible the growth of cities and CIVILIZATIONS. Over the centuries, agricultural production has been increased by IRRIGATION, new varieties of crops and animals, use of agricultural implements and machinery, CROP ROTATION, selective breeding, and genetic engineering.

Agriculture. Growing of crops and raising of LIVESTOCK. Agriculture provides food for human consumption and such products as wool, cotton, and lumber. See also AQUACULTURE.

Air. Colorless, odorless, tasteless, formless mixture of gases that make up the earth's ATMOSPHERE. Comprises almost 78 percent nitrogen and almost 21 percent oxygen, together with small amounts of water vapor, argon, CARBON DIOXIDE, neon, helium, methane, krypton, hydrogen, and other gases, together with minute particles. Air is a synonym for atmosphere; some writers describe the earth's atmosphere as an ocean of air.

Air current. Air currents are caused by differential heating of the earth's surface, which causes heated air to rise. This causes WINDS at the surface as well as higher in the earth's ATMOSPHERE.

Air drainage. Flow of cold, dense air down slopes in response to GRAVITY.

Air mass. Large body of air with distinctive homogeneous characteristics of TEMPERATURE, HUMIDITY, and stability. It forms when air remains stationary over a source REGION for a period of time, taking on the conditions of that region. An air mass can extend over a million square miles with a depth of more than a mile. Air masses are classified according to moisture content (*m* for maritime or *c* for continental) and temperature (*A* for ARCTIC, *P* for polar, *T* for tropical, or *E* for equatorial). The air masses affecting North America are mP, cP, and mT. The interaction of AIR masses produces WEATHER. The line along which air masses meet is a FRONT.

Air pollution. Airborne pollution generated from both natural and man-made sources. Natural sources include pollen from plants, gases and PARTICULATE MATTER from VOLCANOES, and windblown DUST. Artificial sources include industrial and automobile emissions and airborne particles associated with human-induced ABRASION.

Air pressure. See ATMOSPHERIC PRESSURE.

Albedo. Measure of the reflective properties of a surface; the ratio of reflected ENERGY (INSOLATION) to the total incoming energy, expressed as a percentage. The albedo of Earth is 33 percent.

Alberta Clipper. Cold STORM that forms as a low to the east of the Rockies, over Alberta, Canada, and moves rapidly southeast. It brings cold TEMPERATURES, PRECIPITATION, and occasionally heavy SNOW to the northeast United States.

Alienation (land). Land alienation is the appropriation of land from its original owners by a more powerful force. In preindustrial societies, the ownership of agricultural land is of prime importance to subsistence farmers. Colonial governments claimed ownership of the REGIONS they colonized, even though native peoples had lived there for thousands of years. Land in Mexico was alienated by Spain; in Indonesia by the Dutch; and in Australia by the British. The government of the United States alienated most of the lands formerly occupied by NATIVE AMERICANS, leaving them only small "reserves." Some countries have passed laws that attempt to prevent land alienation by foreigners. In Fiji, for example, only Fijians can own or purchase land; the large Indian POPULATION can only lease farmland from Fijian owners.

Alkali flat. Dry LAKEBED in an arid REGION, covered with a layer of SALTS. A well-known example is the Alkali Flat area of White Sands National Monument in New Mexico; it is the bed of a large lake that formed when the GLACIERS were melting. It is covered with a form of gypsum crystals called selenite. This material is blown off the surface into large SAND DUNES. Also called a salina. See also BITTER LAKE.

New Mexico's White Sands National Monument is a well-known example of an alkali flat. (PhotoDisc)

Allogenic sediment. SEDIMENT that originates outside the PLACE where it is finally deposited; SAND, SILT, and CLAY carried by a STREAM into a LAKE are examples.

Alluvial fan. Common LANDFORM at the mouth of a CANYON in arid RE-GIONS. Water flowing in a narrow canyon immediately slows as it leaves the canyon for the wider VALLEY floor, depositing the SEDIMENTS it was transporting. These spread out into a fan shape, usually with a BRAIDED STREAM pattern on its surface. When several alluvial fans grow side by side, they can merge into one continuous sloping surface between the HILLS and the valley. This is known by the Spanish word *bajada*, which means "slope."

Alluvial plain. See FLOODPLAIN.

Alluvial system. Any of various depositional systems, excluding DELTAS, that form from the activity of RIVERS and STREAMS. Much alluvial SEDI-MENT is deposited when rivers top their BANKS and FLOOD the surrounding countryside. Buried alluvial sediments may be important water-bearing RESERVOIRS or may contain PETROLEUM.

Alluvium. Material deposited by running water. This includes not only fertile SOILS, but also CLAY, SILT, or SAND deposits resulting from FLU-VIAL processes. FLOODPLAINS are covered in a thick layer of alluvium.

Alpine. Related to high MOUNTAINS. The alpine OROGENY refers to an episode of mountain formation between 20 and 120 million years ago, which produced the European Alps.

Alpine glacier. Mass of ice and SNOW that moves slowly down from the PEAKS to produce the spectacular LANDFORMS associated with high

Alpine glaciers are characterized by their formation on spectacularly steep mountain slopes.
(PhotoDisc)

MOUNTAIN scenery. Active glaciers may threaten lives and property through catastrophic forward surges and floodwater, or they may be essential sources of MELTWATER in dry areas.

Alternative energy. Renewable forms of ENERGY such as SOLAR, HYDRO-electric, wind, and tidal power; sometimes called sustainable energy. Compare with NONRENEWABLE energy from COAL, OIL, NATURAL GAS.

Altimeter. Instrument for measuring ALTITUDE, or height above the earth's surface, commonly used in airplanes. An altimeter is a type of ANEROID BAROMETER.

Altiplanos. South American term for high PLAINS.

Altitude. Height above the earth's surface, measured from MEAN SEA LEVEL. Pressure decreases regularly with increased altitude, but TEMPERATURE rises or falls depending on the layer of the ATMOSPHERE at which it is measured. The fall of temperature throughout the TROPO-SPHERE (the lowest layer of the atmosphere) leads to ALTITUDINAL ZONATION.

Altitudinal zonation. Existence of different ECOSYSTEMS at various ELEVA-TIONS above SEA LEVEL, due to TEMPERATURE and moisture differences. This is especially pronounced in Central America and South America. The hot and humid COASTAL PLAINS, where bananas and sugarcane thrive, is the *tierra caliente*. From about 2,500 to 6,000 feet (750 to 1,800 meters) is the *tierra templada*; crops grown here include coffee, wheat, and corn, and major cities are situated in this zone. From about 6,000 to 12,000 feet (1,800 to 3,600 meters) is the *tierra fria*; here only hardy crops such as potatoes and barley are grown, and large numbers of animals are kept. From about 12,000 to 15,000 feet (3,600 to 4,500 meters) lies the *tierra helada*, where hardy animals such as sheep and alpaca graze. Above 15,000 feet (4,500 meters) is the frozen *tierra nevada*; no permanent life is possible in the permanent SNOW and ICE FIELDS there.

Altocumulus. Puffy CLOUD masses at a middle ALTITUDE, between 20,000 and 43,000 feet (6,000 to 13,000 meters) above the earth's surface. Sizes and shapes of these clouds vary; their colors are grey and white.

Altostratus. Layers of CLOUD covering a large part of the sky at a middle ALTITUDE, between 20,000 and 43,000 feet (2,000 to 6,000 meters) above the earth's surface. Altostratus clouds may produce continuous rain. Usually a uniform grey or blue grey in color. The SUN may shine weakly through altostratus clouds.

Amerindians. Contraction of "American Indians"; widely accepted term for the native peoples of the Caribbean, Central America, and North America. See also NATIVE AMERICANS.

Anabatic wind. Upslope WIND, blowing up a HILL or MOUNTAIN as the result of strong surface heating of the slopes. Similar to a valley BREEZE.

Andesite. Volcanic IGNEOUS ROCK type intermediate in composition and density between GRANITE and BASALT.

Anemometer. Instrument for measuring WIND speed or wind velocity, consisting of a set of cups or cones that rotate as the wind blows into them. See also ALTIMETER; BAROMETER.

Aneroid barometer. Sealed, partially evacuated box connected to a needle and dial, used to measure changes in ATMOSPHERIC PRESSURE. See also ANEMOMETER; BAROMETER.

Angle of repose. Maximum angle of steepness that a pile of loose materials such as SAND or ROCK can assume and remain stable; the angle varies with the size, shape, moisture, and angularity of the material.

Animism. Belief that natural features, such as LANDFORMS, plants, and animals, possess a spirit. This spirit can intervene in the real world to bring good or bad fortune to a person. In religious terms, gods dwell in certain sacred PLACES, or take the shape of a certain animal. Alternatively, a person's soul or spirit may pass into a certain animal or landform after death. An animistic view of nature is still found in many African CULTURES. The Australian Aborigines have an animist view of the land and its inhabitants. Although animism is sometimes characterized as primitive, it is thought that all RELIGIONS began as animist beliefs and customs. Animism is regarded favorably by conservationists as an environmentally responsible philosophy.

Anorthosite. IGNEOUS ROCK, solidified from the molten state, consisting mostly of FELDSPAR. Coarse-grained, INTRUSIVE igneous rocks composed principally of plagioclase feldspar, anorthosites are useful for what they reveal about the early crustal evolution of the earth, and they are the source of several economic commodities.

Antarctic. Relating to the REGION south of the Antarctic Circle, extending from 66.5 DEGREES south to the South Pole at 90 degrees south. The CONTINENT of Antarctica is located there. The international Antarctic Treaty allows for scientific research in Antarctica by several NATIONS, but prohibits military use.

Antecedent river. STREAM that was flowing before the land was uplifted and was able to erode at the pace of UPLIFT, thus creating a deep CANYON. Most deep canyons are attributed to antecedent rivers. In the Davisian CYCLE OF EROSION, this process was called REJUVENATION.

Anthropocentric. Regarding humanity as the center or most important consideration. An anthropocentric view of nature holds that all plants and animals exist primarily for human use and benefit.

Anthropogeography. Branch of GEOGRAPHY founded in the late nineteenth century by German geographer Friedrich Ratzel. The field is closely related to human ECOLOGY—the study of humans, their DISTRIBUTION over the earth, and their interaction with their physical ENVIRONMENT.

Anticline. Area where land has been UPFOLDED symmetrically. Its center contains stratigraphically older ROCKS. See also SYNCLINE.

Anticyclone. High-pressure system of rotating WINDS, descending and di-

Anticline on the banks of the Potomac River. (U.S. Geological Survey)

verging, shown on a WEATHER chart by a series of closed ISOBARS, with a high in the center. In the NORTHERN HEMISPHERE, the rotation is CLOCKWISE; in the SOUTHERN HEMISPHERE, the rotation is COUNTER-CLOCKWISE. An anticyclone brings warm weather. See also CYCLONE.

Antidune. Undulatory upstream-moving bed form produced in free-surface flow of water over a SAND bed in a certain RANGE of high flow speeds and shallow flow depths.

Antipodes. TEMPERATE ZONE of the SOUTHERN HEMISPHERE. The term is now usually applied to the countries of Australia and New Zealand. The ancient Greeks believed that if humans existed there, they must walk upside down. This idea was supported by the Christian Church in the Middle Ages.

Antitrade winds. WINDS in the upper ATMOSPHERE, or GEOSTROPHIC winds, that blow in the opposite direction to the TRADE WINDS. Anti-trade winds blow toward the northeast in the NORTHERN HEMISPHERE and toward the southeast in the SOUTHERN HEMISPHERE.

Aperiodic. Irregularly occurring interval, such as found in most WEATHER CYCLES, rendering them virtually unpredictable.

Aphelion. Point in the earth's 365-DAY REVOLUTION when it is at its greatest distance from the SUN. This is caused by Earth's elliptical

ORBIT around the Sun. The distance at aphelion is 94,555,000 miles (152,171,500 km.) and usually falls on July 4. The opposite of PERIHELION.

Aplite. Light-colored, sugary-textured granitic ROCK generally found as small, late-stage VEINS in GRANITES of normal TEXTURE; in pegmatites, aplites usually form thin marginal selvages against the country rock but may also occur as major lenses in the pegmatite interior.

Apogee. Point in the MOON's ORBIT when it is most distant from the center of the earth.

Aposelene. Earth's farthest point from the MOON.

Apparent solar time. Time shown on a sundial; also known as apparent time. Because the elliptical path of the earth around the SUN causes the apparent solar DAY to vary, there can be a difference of almost four minutes in the length of a day. When accurate CLOCKS and watches were developed in the seventeenth century, mean solar time was adopted instead of apparent solar time.

Aquaculture. Commercial raising and harvesting of food such as fish, shellfish, or seaweed in artificial ponds or in selected portions of RIVERS or coastal waters.

Aqueduct. Artificial conduit for carrying water. See also CANAL.

Ruins of Roman aqueduct at Carthage in what is now Tunis. (Arkent Archive)

Aquifer. Underground body of POROUS ROCK that contains water and allows water PERCOLATION through it. The largest aquifer in the United States is the Ogallala Aquifer, which extends south from South Dakota to Texas.

Arable land. Land that is suitable for farming. A measure of the productivity of arable land is the PHYSIOLOGIC DENSITY.

Archaeology. Science that investigates the human past through its material remains, such as pots, tools, buildings, and human remains.

Archipelago. Group of ISLANDS located close together; an island chain. Indonesia, comprising thirteen thousand islands, is the world's largest archipelago.

Arctic. Relating to the REGION north of the Arctic Circle, extending from 66.5 DEGREES north to the North Pole at 90 degrees north. Comes from the Greek word meaning bear, because the Arctic region is located under the constellation Ursa Major or Big Bear.

Arête. Serrated or saw-toothed ridge, produced in glaciated MOUNTAIN areas by CIRQUES eroding on either side of a RIDGE or mountain RANGE. From the French word for knife-edge.

Arête and talus slopes in Colorado's San Juan Mountains. (U.S. Geological Survey)

Aridity. Lack of PRECIPITATION. No trees or woody plants can grow in arid REGIONS, and VEGETATION is small and sparse.

Arroyo. Spanish word for a dry STREAMBED in an arid area. Called a WADI in Arabic and a WASH in English.

Artesian well. WELL from which GROUNDWATER flows without mechanical pumping, because the water comes from a CONFINED AQUIFER, and is therefore under pressure. The Great Artesian Basin of Australia has hundreds of artesian wells, called BORES, that provide drinking water for sheep and cattle. The name comes from the Artois REGION of France, where the phenomenon is common. A subartesian well is sunk into an UNCONFINED AQUIFER and requires a pump to raise water to the surface.

Aseismic. Lacking EARTHQUAKE activity.

Ash. Fine-grained pyroclastic material less than 2 millimeters in diameter, ejected from an erupting VOLCANO. See also VOLCANIC ASH.

Volcano spewing ash into the atmosphere. (PhotoDisc)

Ash flow. Density current composed of a highly heated mixture of volcanic gases and ASH, which travels down the flanks of a VOLCANO or along the ground surface.

Assimilation. Absorption of one cultural group into a larger or dominant cultural group through the former group's adoption of cultural traits such as LANGUAGE, clothing, and customs from the latter group. Immigrants to the United States in the nineteenth century were assimilated as part of the "melting pot"; in the late twentieth century, in contrast, retaining the original CULTURE was encouraged, as part of MULTICULTURALISM.

Asteroid. Small PLANET or celestial body made of ROCK that moves around the SUN, usually in an ORBIT between Mars and Jupiter, where there are approximately forty-five thousand asteroids.

Asteroid belt. REGION between the ORBITS of Mars and Jupiter containing the majority of ASTEROIDS.

Asthenosphere. Part of the earth's UPPER MANTLE, beneath the LITHO-SPHERE, in which PLATE movement takes place. Also known as the low-velocity zone.

Astrobleme. Remnant of a large IMPACT CRATER on Earth.

Astrolabe. Medieval instrument used to measure the ALTITUDE of celestial bodies, especially the SUN.

Astronomical unit (AU). Unit of measure used by astronomers that is equivalent to the average distance from the SUN to Earth (93 million miles/150 million km.).

Atlas. Book of MAPS, often accompanied by CHARTS, tables, and illustrations. Named after the figure in Greek mythology who was condemned by Zeus to carry the heavens on his shoulders.

Atmosphere. Mixture of gases surrounding the earth. The atmosphere is thinner at the POLES than at the EQUATOR and varies with the SEASON, but averages 300 miles (480 km.) above the earth's surface. Beyond the atmosphere is the EXOSPHERE. The modern atmosphere differs from that of four billion years ago in that it contains oxygen. The modern atmosphere is 78 percent nitrogen by volume and almost 21 percent oxygen. Other atmospheric gases include argon, CARBON DIOX-IDE, neon, helium, methane, and krypton. Small amounts of OZONE, nitrous oxide, hydrogen, and xenon also occur, as do variable amounts of water vapor and PARTICULATE MATTER. Some scientists believe that human activities are changing the atmosphere so rapidly that one can speak of an ANTHROPOGENIC atmosphere. GLOBAL WARMING is thought by many to be a result of anthropogenic change, especially the increase in carbon dioxide.

Atmospheric pressure. Weight of the earth's ATMOSPHERE, equally distributed over earth's surface and pressing down as a result of GRAVITY. On average, the atmosphere has a force of 14.7 pounds per square inch (1 kilogram per centimeter) squared at SEA LEVEL, also expressed as 1013.2 millibars. Variations in atmospheric pressure, high or low, cause WINDS and WEATHER changes that affect CLIMATE. Pressure decreases rapidly with ALTITUDE or distance from the surface: Half of the total atmosphere is found below 18,000 feet (5,500 meters); more than 99 percent of the atmosphere is within 30 miles (50 km.) of the surface. Atmospheric pressure is measured with a BAROMETER.

Atoll. Ring-shaped growth of CORAL REEF, with a LAGOON in the middle. Charles Darwin, who observed many Pacific atolls during his voyage on the *Beagle* in the nineteenth century, suggested that they were created from FRINGING REEFS around volcanic ISLANDS. As such islands sank beneath the water (or as SEA LEVELS rose), the coral continued growing upward. SAND resting atop an atoll enables plants to grow, and small human societies have arisen on some atolls. The world's largest atoll, Kwajalein in the Marshall Islands, measures about 40 by 18 miles (65 by 30 km.), but perhaps the most famous atoll is Bikini Atoll—the SITE of nuclear-bomb testing during the 1950's.

Atomic clock. Extremely precise timekeeper that uses the vibration or natural frequency of cesium atoms to measure time. The first atomic clock was built in the United States in 1949; the first using cesium in 1952. High-accuracy timekeeping became increasingly important with the advent of telecommunications, especially for the GLOBAL POSITIONING SYSTEM. World time, also called UTC or COORDINATED UNIVERSAL TIME, is now measured using standard clocks, which are all atomic clocks. The world's most accurate clock, the Cesium Fountain Clock NST F-1, went into operation at Boulder, Colorado, in late 1999. Its accuracy is such that it should not gain or lose a second if it were keep running for twenty million years.

Aurora. Glowing and shimmering displays of colored lights in the upper ATMOSPHERE, caused by interaction of the SOLAR WIND and the charged particles of the IONOSPHERE. Auroras occur at high LATITUDES. Near the North Pole they are called aurora borealis or northern lights; near the South Pole, aurora australis or southern lights.

Aurora borealis and bright moon, viewed from earth orbit. (Corbis)

Austral. Referring to an object or occurrence that is located in the SOUTHERN HEMISPHERE or related to Australia. Compare with BOREAL.

Australopithecines. Erect-walking early human ancestors with a cranial capacity and body size within the RANGE of modern apes rather than of humans.

Autumnal equinox. See EQUINOX.

Avalanche. Mass of SNOW and ice falling suddenly down a MOUNTAIN slope, often taking with it earth, ROCKS, and trees.

Axis of the earth. Imaginary line passing through the center of the earth from the North Pole to the South Pole. The earth rotates on its axis once in every twenty-four hours, in a COUNTERCLOCKWISE direction if viewed from above the North Pole, or in a west-to-east direction if viewed from above the EQUATOR. As a result, the SUN appears to rise in the east and set in the west.

Azimuth. DEGREES of arc measured CLOCKWISE from the north.

Azimuthal projection. Projection that can be visualized by imagining a sheet of paper resting at a point on the surface of a center-lit globe, so that the outlines of CONTINENTS are projected onto the paper. Usually this projection is used for MAPS of the ARCTIC or ANTARCTIC regions. There is no distortion at the point of tangency or contact, but increased distortion of both shape and area with distance away from the center of the map. Also called a plane projection.

B horizon. SOIL layer just beneath the TOPSOIL.

Backswamp. See BAYOU.

Bajada. See ALLUVIAL FAN.

Bank. Elevated area of land beneath the surface of the OCEAN. The term is also used for elevated ground lining a body of water.

Bar (climate). Measure of ATMOSPHERIC PRESSURE per unit surface area of one million dynes per square centimeter. Millibars (thousandths of a bar) are the MEASUREMENT used in the United States. Other countries use kilopascals (kPa); one kilopascal is ten millibars.

Bar (land). RIDGE or long deposit of SAND or gravel formed by DEPOSITION in a RIVER or at the COAST. Offshore bars and baymouth bars are common coastal features.

Barogram. Chart or record made by a BAROGRAPH.

Barograph. BAROMETER that is equipped with a device to provide a continuous record of ATMOSPHERIC PRESSURE.

Barometer. Instrument used for measuring ATMOSPHERIC PRESSURE. In the seventeenth century, Evangelista Torricelli devised the first barometer—a glass tube sealed at one end, filled with mercury, and upended into a bowl of mercury. He noticed how the height of the mercury column changed and realized this was a result of the pressure of air on the mercury in the bowl. Early MEASUREMENTS of atmospheric pressure were, therefore, expressed as centimeters of mercury, with average pressure at SEA LEVEL being 29.92 inches (760 millimeters). This cumbersome barometer was replaced with the ANEROID BAROMETER—a sealed and partially evacuated box connected to a needle and dial, which shows changes in atmospheric pressure. See also ALTIMETER.

Barrier island. Long chain of SAND islands that forms offshore, close to the COAST. LAGOONS or shallower MARSHES separate the barrier islands from the mainland. Such LOCATIONS are hazardous for SETTLEMENTS because they are easily swept away in STORMS and HURRICANES. In the United States, barrier islands extend from the Texas coast to the Outer Banks of North Carolina and on to Long Island. Cape Hatteras is part of the barrier islands, being composed of sand and not part of the mainland.

Barysphere. Dense, heavy CORE of the earth.

Basalt. IGNEOUS EXTRUSIVE ROCK formed when LAVA cools; often black in color. Sometimes basalt occurs in tall hexagonal columns, such as the Giant's Causeway in Ireland, or the Devils Postpile at Mammoth, California.

Basalt cliffs in Yellowstone National Park. An igneous extrusive rock formed when lava cools, basalt is typically black in color. (Corbis)

Base flow. Natural flow of GROUNDWATER into a RIVER, which commonly maintains the flow of PERENNIAL STREAMS during dry seasons.

Base level. Level below which a STREAM cannot erode its bed or VALLEY. For most RIVERS, the ULTIMATE BASE LEVEL is MEAN SEA LEVEL. For rivers that flow into a LAKE, there is a local base level, which is the level of the lake. A section of resistant ROCK might provide a local base level, but this would change through EROSION over time. The base-level concept was developed by John Wesley Powell in the nineteenth century after exploring the Colorado River and Grand Canyon.

Basement. Crystalline, usually PRECAMBRIAN, IGNEOUS and METAMORPHIC ROCKS that occur beneath the SEDIMENTARY ROCK on the CONTINENTS.

Basin. REGION drained by a RIVER system, including all of its tributaries. See also DRAINAGE BASIN.

Basin order. Approximate measure of the size of a STREAM BASIN, based on a numbering scheme applied to RIVER channels as they join together in their progress downstream.

Batholith. Large LANDFORM produced by IGNEOUS INTRUSION, composed of CRYSTALLINE ROCK, such as GRANITE; a large PLUTON with a surface area greater than 40 square miles (100 sq. km.). Most mountain RANGES have a batholith underneath.

Fry Creek Batholith in British Columbia. (Geological Survey of Canada)

Bathymetric contour. Line on a MAP of the OCEAN floor that connects points of equal depth.

Bauxite. Principal ORE from which aluminum is obtained. Usually found in the wet TROPICS, although the name comes from a REGION of France.

One of the largest and finest natural harbors in the world is San Francisco Bay, an immense inlet protected from the Pacific Ocean by the San Francisco and Marin peninsulas. (PhotoDisc)

Bay. Part of a SEA or OCEAN partially enclosed by land, such as the Bay of Biscay. "Bay" is not a precise term, but it is usually applied to ocean IN-LETS smaller than a GULF.

Bayou. Low-lying, swampy area near a RIVER. After a river on a FLOOD-PLAIN overflows, some water remains, creating a marshy area on either side of the STREAM beyond the NATURAL LEVEES. "Bayou" is a Cajun word. Also called a backswamp.

Beach. Part of a COAST where SEDIMENT has accumulated and is moved by

Honolulu's Waikiki Beach on the Hawaiian island of Oahu is one of the most popular sandy beaches in the world. (PhotoDisc)

WAVES and CURRENTS. The beach zone extends from above the high-tide level to below the low-tide level. Most beaches are covered in SAND; when rounded ROCKS, PEBBLES, or cobbles cover a beach, it is called a shingle beach.

Beaufort scale. SCALE that measures WIND force, expressed in numbers from 0 to 12. The original Beaufort scale was based on descriptions of the state of the SEA. It was adapted to land conditions, using descriptions of chimney smoke, leaves of trees, and similar factors. The scale was devised in the early nineteenth century by Sir Francis Beaufort, a British naval officer.

Bedrock. Solid ROCK covered by SOIL, which is part of the earth's CRUST. When the covering material is removed and the rock exposed at the surface, it is called an OUTCROP.

Belt. Geographical REGION that is distinctive in some way.

Bergeron process. PRECIPITATION formation in COLD CLOUDS whereby ice crystals grow at the expense of supercooled water droplets.

Bight. Wide or open BAY formed by a curve in the COASTLINE, such as the Great Australian Bight.

Billabong. Australian term for a waterhole.

Typical Australian billabong—the kind of waterhole at which the Swagman camped in Australia's unofficial national anthem, "Waltzing Mathilda." (Ray Sumner)

Biodiversity. Measure of the variety of life occupying a particular ECO-SYSTEM.

Biogenic sediment. SEDIMENT particles formed from skeletons or shells of microscopic plants and animals living in seawater.

Biogeography. Study of the worldwide DISTRIBUTION of ECOSYSTEMS (plants and animals); also the study of changes in these distributions over time.

Biome. Large ECOSYSTEM on a continental scale; a terrestrial ecosystem. Specific combinations of plants and animals, known as communities, live in each biome.

Biosphere. Parts of Earth in which life exists; includes the lower part of the ATMOSPHERE, the HYDROSPHERE, and the upper LITHOSPHERE (Earth's CRUST). The term is also applied to the complex totality of plant and animal life on Earth.

Biostratigraphy. Identification and organization of STRATA based on their FOSSIL content and the use of fossils in stratigraphic correlation.

Biotechnology. Range of scientific techniques using living tissue, seeds, or organisms to make improved varieties of crops or animals, thereby increasing food production. Biotechnology has led to high-yield, pest-resistant, and DROUGHT-tolerant crops. However, many of the innovations of biotechnology, such as cloning, are controversial. Another issue is that biotechnological research is carried out by MULTINATIONAL CORPORATIONS, whose products are not readily available to the poor RURAL communities whose need for food is greatest. Another fear associated with biotechnology is the security of the world food supply as private firms gain increased control over food production, rather than governments of individual countries.

Birth rate. Annual number of births per one thousand people, in any given POPULATION under study. The birth rate for the United States was fifteen (per thousand per year) at the end of the twentieth century. "Birth rate" is a shortened term for CRUDE BIRTH RATE.

Bitter lake. Saline or BRACKISH LAKE in an arid area, which may dry up in the summer or in periods of DROUGHT. The water is not suitable for drinking. An example is the Bitter Lake Wildlife Refuge in New Mexico, which provides a resting place for huge numbers of migratory birds each year, including Canadian snow geese. Another name for this feature is "salina." See also ALKALI FLAT.

Blizzard. Intense cold STORM in which WINDS reach speeds of at least 35 miles (56 km.) per hour, TEMPERATURES drop below 20 DEGREES Fahrenheit (−7 degrees Celsius), visibility falls below 820 feet (250 meters), and all these conditions last a minimum of three hours. Snowfall often accompanies a blizzard, but this is not a necessary condition; much of the SNOW is simply driven by the strong winds.

Block lava. LAVA flows whose surfaces are composed of large, angular blocks; these blocks are generally larger than those of AA flows and have smooth, not jagged, faces.

Block mountain. MOUNTAIN or mountain RANGE with one side having a gentle slope to the crest, while the other slope, which is the exposed FAULT SCARP, is quite steep. It is formed when a large block of the earth's CRUST is thrust upward on one side only, while the opposite side remains in place. The Sierra Nevada in California are a good example of block mountains. Also known as fault-block mountain.

Blowhole. SEA CAVE or tunnel formed on some rocky, rugged COASTLINES. The pressure of the seawater rushing into the opening can force a jet of seawater to rise or spout through an opening in the roof of the cave. Blowholes are found in Scotland, Tasmania, and Mexico, and on the Hawaiian ISLANDS of Kauai and Maui.

Bluff. Steep slope that marks the farthest edge of a FLOODPLAIN.

Body wave. SEISMIC WAVE that propagates interior to a body; there are two kinds, P WAVES and S WAVES, that travel through the earth, reflecting and refracting off the several layered boundaries within the earth.

Bog. Damp, spongy ground surface covered with decayed or decaying VEGETATION. Bogs usually are formed in cool CLIMATES through the in-filling, or silting up, of a LAKE. Moss and other plants grow outward toward the edge of the lake, which gradually becomes shallower, until the surface is completely covered. Bogs also can form on cold, damp MOUNTAIN surfaces. Many bogs are filled with PEAT.

Bora. Strong, cold, squally downslope WIND on the Dalmatian COAST of Yugoslavia in winter. A KATABATIC WIND.

Border. Technically, the area on either side of a BOUNDARY. The term commonly is used instead of "boundary" to mean the imaginary line separating one COUNTRY from another. The boundary between the United States and Canada, along the forty-ninth PARALLEL north, is the world's longest undefended border.

Bore. Standing WAVE, or wall, of water created in a narrow ESTUARY when the strong incoming, or FLOOD, TIDE meets the RIVER water flowing outward; it moves upstream with the advancing tide, and downstream with the EBB TIDE. South America's Amazon River and Asia's Mekong River have large bores. In North America, the bore in the Bay of Fundy is visited by many tourists each year. Its St. Andrew's wharf is designed to handle changes in water level of as much as 53 feet (15 meters) in one DAY.

Boreal. Alluding to an item or event that is in the NORTHERN HEMISPHERE. Compare with AUSTRAL.

Boreal forest. FORESTS found at LATITUDES above 50 DEGREES north in North America, Europe, and Asia. Because of the intense cold, the trees are needleleaf species, such as spruce and fir. Unlike temperate or tropical forests, boreal forests have little undergrowth; instead, the forest floor is covered with mosses and lichen, which also grow on the tree trunks. Many animals live in the boreal forest, surviving the cold either through MIGRATION or hibernation.

Bottom current. Deep-sea current that flows parallel to BATHYMETRIC CONTOURS.

Bottom-water mass. Body of water at the deepest part of the OCEAN identified by similar patterns of SALINITY and TEMPERATURE.

Boundary. Imaginary line that separates political units from one another. A boundary can be a straight line or a geometric boundary, such as the forty-ninth PARALLEL separating Canada and the United States; other boundaries follow RIVERS, mountain RANGES, or other natural features. People sometimes use the term "BORDER" when speaking about a boundary.

Bourne. English term for a small STREAM or BROOK. Similar to the Scottish word "burn."

Brackish water. Water with SALT content between that of SALT WATER and FRESH WATER; it is common in arid areas on the surface, in coastal MARSHES, and in salt-contaminated GROUNDWATER.

Brae. Scottish word for the hillside or BANKS of a RIVER.

Braided stream. STREAM having a CHANNEL consisting of a maze of interconnected small channels within a broader STREAMBED. Braiding occurs when the stream's load exceeds its capacity, usually because of reduced flow.

Breaker. WAVE that becomes oversteepened as it approaches the SHORE, reaching a point at which it cannot maintain its vertical shape. It then breaks, and the water washes toward the shore.

A breaker is a wave that becomes oversteepened as it approaches the shore, reaching a point at which it cannot maintain its vertical shape. It then breaks, and the water washes toward the shore. (PhotoDisc)

Breakwater. Large structure, usually of ROCK, built offshore and parallel to the COAST, to absorb WAVE ENERGY and thus protect the SHORE. Between the breakwater and the shore is an area of calm water, often used as a boat anchorage or HARBOR. A similar but smaller structure is a seawall.

Breccia. See CONGLOMERATE.

Breeze. Gentle WIND with a speed of 4 to 31 miles (6 to 50 km.) per hour.

On the BEAUFORT SCALE, the numbers 2 through 6 represent breezes of increasing strength.

Bridge. Physical structure spanning a RIVER, roadway, or other GAP or obstacle. Artificially created bridges are usually used to provide passageways.

Brine. Usually warm, highly saline seawater containing calcium, sodium, potassium, chlorine, and other small amounts of free ions.

Brook. Natural STREAM of water, smaller than a RIVER, issuing from a SPRING.

Bush. Relatively small plant with leafy foliage on several stems that branch close to the ground. The word "shrub" is also used. In Australia, "bush" is a term for any unspecified nonurban area.

Butte. Flat-topped HILL, smaller than a MESA, found in arid REGIONS.

Caldera. Large circular depression with steep sides, formed when a VOLCANO explodes, blowing away its top. The ERUPTION of Mount St. Helens produced a caldera. Crater Lake in Oregon is a caldera that has filled with water. From the Spanish word for kettle.

Caldera. (PhotoDisc)

Calendar. System of dividing time into years, months, and DAYS, based on observations of the SUN, MOON, and stars. The basic unit is the day, which now is measured from one midnight to the next, but often was measured from one DAWN to the next in ancient times. The seven-day week is based on the approximate length of each of the four phases of the Moon. The Julian calendar, with a year length of 365 and one-quarter days, was introduced to the Western world by Julius Caesar in 46 B.C. (The month July commemorates Caesar.) The Julian calendar year was too long by about eleven minutes, so by the sixteenth century the calendar had become out of phase with the SEASONS and religious holidays were falling inappropriately. Pope Gregory XIII, advised by astronomers, determined to omit ten days from the calendar to correct the errors. The Gregorian calendar developed as a result gradually was adopted in other European countries. One of the last countries to adopt it was Russia, in 1918. Ancient peoples had different calendars, and RELIGIONS other than Christianity use different calendars—for example, the year 2000 is 5760 in the Jewish calendar and 1378 in the Muslim calendar.

Calms of Cancer. Subtropical BELT of high pressure and light WINDS, located over the OCEAN near 25 DEGREES north LATITUDE. Also known as the HORSE LATITUDES.

Calms of Capricorn. Subtropical BELT of high pressure and light WINDS, located over the OCEAN near 25 DEGREES south LATITUDE.

Calving. Loss of glacial mass when GLACIERS reach the SEA and large blocks of ice break off, forming ICEBERGS.

Cambrian period. PERIOD from about 570 to 505 million years ago, marked by the appearance of hard-shelled organisms.

Canal. Artificial waterway constructed to shorten the route between two PLACES. Often a canal is cut through an ISTHMUS, as with the Suez Canal and the Panama Canal. Canals also are built to connect two RIVERS, such as the Grand Canal of China, or to bring IRRIGATION water to an arid REGION.

Cancer, tropic of. PARALLEL of LATITUDE at 23.5 DEGREES north; this line is the latitude farthest north on the earth where the noon SUN is ever directly overhead. The REGION between it and the tropic of CAPRICORN is known as the TROPICS.

Canyon. Steep-sided STREAM VALLEY or GORGE in an arid REGION. The

A canyon is steep-sided stream valley or gorge. (PhotoDisc)

most famous North American canyon is the Grand Canyon in the southwestern United States.

Cape. Point of land that protrudes beyond the nearby COAST into the SEA or a LAKE. See also HEADLAND.

Capillary water. Water held in the upper part of the SOIL by surface tension of water around the soil particles. See also SOIL MOISTURE.

Capital. CITY that is the seat of a regional or national government.

Capitol. Building that houses a government legislature.

Capricorn, tropic of. Line of LATITUDE at 23.5 DEGREES south; this line is the latitude farthest south on the earth where the noon SUN is ever directly overhead. The REGION between it and the tropic of CANCER is known as the TROPICS.

Carbon cycle. Changes that carbon undergoes in the BIOSPHERE, starting with the conversion by PHOTOSYNTHESIS of atmospheric CARBON DIOXIDE into biomass and its return to a gaseous form during RESPIRATION and decay processes.

Carbon dating. Method employed by physicists to determine the age of organic matter—such as a piece of wood or animal tissue—to determine the age of an archaeological or paleontological SITE. The method works on the principle that the amount of radioactive carbon in living matter diminishes at a steady and measurable rate after the matter dies. Technique is also known as carbon-14 dating, after the radioactive carbon-14 isotope it uses. Also known as radiocarbon dating.

Carbon dioxide. Gas that occurs naturally in the earth's modern ATMOSPHERE, contributing 0.036 percent by volume at the end of the twentieth century. It is produced naturally through RESPIRATION of living organisms and is part of the CARBON CYCLE. The amount of carbon dioxide in the earth's atmosphere has increased over the last two centuries, from 0.028 percent in 1774. This is largely as a result of burning FOSSIL FUELS for ENERGY, which began on a large scale with the INDUSTRIAL REVOLUTION. DEFORESTATION also has contributed to the increased level of carbon dioxide. Because carbon dioxide reflects EARTH RADIATION back to the surface, it is believed to play a large role in GLOBAL WARMING. The United States is the world's largest user of energy and, therefore, the largest producer of carbon dioxide.

Carbonates. Large group of MINERALS consisting of a carbonate anion (three oxygen atoms bonded to one carbon atom, with a residual charge of two) and a variety of cations, including calcium, magnesium, and iron.

Carboniferous period. Fifth of the six PERIODS in the PALEOZOIC ERA; it preceded the PERMIAN PERIOD and spanned a period of 320 to 286 million years ago.

Cardinal points. Four main points of the COMPASS: north, south, east, and west.

Carnivore. Animal that eats mainly flesh. See also FOOD CHAIN.

Carrying capacity. Number of animals that a given area of land can support, without additional feed being necessary. Lush GRASSLAND may have a carrying capacity of twenty sheep per acre, while more arid, SEMIDESERT land may support only two sheep per acre. The term sometimes is used to refer to the number of humans who can be supported in a given area.

Cartography. Specialized science of producing MAPS or CHARTS, which draws on mathematics and art as well as geography. Computer-based cartography developed rapidly at the end of the twentieth century.

Cascade. Series of small WATERFALLS in a rocky part of a STREAMBED.

A cascade is a series of small waterfalls in a rocky part of a stream bed. (PhotoDisc)

Cataract. Large WATERFALL. The Nile River in Africa was impassable to shipping for centuries because of several cataracts.

Catastrophism. Theory, popular in the eighteenth and nineteenth centuries, that explained the shape of LANDFORMS and CONTINENTS and the EXTINCTION of species as the results of intense or catastrophic events.

The biblical FLOOD of Noah was one such event, which supposedly explained many extinctions. Catastrophism is linked closely to the belief that the earth is only about six thousand years old, and therefore tremendous forces must have acted swiftly to create present LANDSCAPES. An alternative or contrasting theory is UNIFORMITARIANISM.

Catchment basin. Area of land receiving the PRECIPITATION that flows into a STREAM. Also called catchment or catchment area.

Causeway. Elevated path or road above water or marshy ground.

Cave. Natural underground opening. Caves commonly form in areas of LIMESTONE ROCK, through SOLUTION of the rock by water. The world's largest system of interconnected caves is in Mammoth Cave National Park in Kentucky. The world's largest single cave is in Sarawak, on the ISLAND of Borneo; the deepest cave is in France. People who explore caves are called speleologists. See also KARST.

Cay. Small ISLANDS or ISLETS of SAND above CORAL REEFS. The term "cay" is used in countries such as Australia; in the United States, they are called KEYS, for example, the Florida Keys.

Celsius scale. TEMPERATURE SCALE devised by Anders Celsius, in which the melting point of ice at SEA LEVEL is zero DEGREES and the boiling point of water at sea level is one hundred degrees. Most countries except the United States use the Celsius scale for temperature MEASUREMENT. The Celsius scale formerly was called the centigrade scale. See also FAHRENHEIT SCALE.

Cenozoic era. PERIOD of geologic time from about 65 million years ago to the present. The youngest of the three PHANEROZOIC EONS, it encompasses two geologic periods, the TERTIARY (older) and the QUATERNARY. Through study of the GEOLOGIC RECORD from this era, scientists are able to distinguish between environmental changes caused by a normal progression of geologic phenomena and those changes that are related to human activity.

Census. Official counting of the POPULATION of a COUNTRY to obtain DEMOGRAPHIC data. The United States takes census every ten years.

Centigrade scale. See CELSIUS SCALE.

Central place theory. Theory that explains why some SETTLEMENTS remain small while others grow to be middle-sized TOWNS, and a few become large cities or METROPOLISES. The explanation is based on the provision of goods and services and how far people will travel to acquire these. The German geographer Walter Christaller developed this theory in the 1930's.

Central places. SETTLEMENTS where goods and services are available to consumers from the surrounding area or REGION. If a PLACE offers few services, the POPULATION will be correspondingly small. From another point of view, small places offer certain essential services, such as a gas station, a convenience store, restaurants, and an elementary school. A larger place offers the previous services, plus perhaps a supermarket, cinema, high school, and post office. Central places are organized hierarchically. There are large numbers of small settlements, relatively closely spaced; there are fewer large cities, located farther apart.

Centrality. Measure of the number of functions, or services, offered by any CITY in a hierarchy of cities within a COUNTRY or a REGION. See also CENTRAL PLACE THEORY.

Centrifugal forces. Forces that divide a COUNTRY. Cultural differences, such as two different LANGUAGES or two different RELIGIONS, are important centrifugal forces. The independence movement in Quebec is a good example of the operation of centrifugal forces. When centrifugal forces outweigh CENTRIPETAL FORCES, a country can break up into smaller units. This process is called DEVOLUTION.

Centripetal forces. Forces that unite a COUNTRY. Cultural characteristics, such as a common LANGUAGE or a single RELIGION, are important centripetal forces. New countries create symbols of unity, such as a national flag and national anthem. A powerful leader can be a strong centripetal force, as can war against a common enemy.

CFC. See CHLOROFLUOROCARBONS.

Chain, island. See ARCHIPELAGO.

Chain, mountain. Another term for mountain RANGE.

Chalk. Naturally occurring sedimentary deposit of soft calcium carbonate. The White Cliffs of Dover are a well-known chalk LANDFORM; EROSION is occurring quickly along that part of the English COAST.

Channel. STREAM channels carry water that falls as PRECIPITATION, or comes from melted SNOW, from one PLACE to another, with the water moving downchannel as a result of GRAVITY. A stream channel changes in width and depth because the volume and speed of the water varies. Channels are usually sinuous, rather than straight. On FLOODPLAINS, the channel becomes a series of MEANDERS. BRAIDED STREAM patterns occur with low flow and high SEDIMENT transport. In arid areas, dry STREAMBEDS (WADIS) are common.

Chaparral. Distinctive shrubland VEGETATION that grows around the SHORES of the Mediterranean Sea (where it is called *maquis*), and in areas of MEDITERRANEAN CLIMATE in California, at the southern tip of South Africa, in central Chile, and in two small REGIONS of western and southern Australia. To adapt to the extreme conditions of a long dry summer and wet winter, plants in this BIOME have small leaves, sometimes with a wax-like coating, and usually have deep root systems. Chaparral regenerates quickly after fire, which is frequent in the summer in the Mediterranean climate.

Chart. MAP indicating dangerous areas, used for NAVIGATION by air and SEA. An aeronautical chart shows MOUNTAINS, towers, and airstrips; a nautical chart shows lighthouses, REEFS, and water depths.

Chemical farming. Application of artificial FERTILIZERS to the SOIL and the use of chemical products such as insecticides, fungicides, and her-

The invention of heavier-than-air flight in the early twentieth century made possible efficient large-scale application of chemical fertilizers with the use of airplanes, popularly known as "crop dusters." (PhotoDisc)

bicides to ensure crop success. Chemical farming is practiced mainly in high-income countries, because the cost of the chemical products is high. Farmers in low-income economies rely more on natural organic fertilizers such as animal waste.

Chemical weathering. Chemical decomposition of solid ROCK by processes involving water that change its original materials into new chemical combinations.

Chinook. Warm WIND that melts SNOWS on the Canadian PRAIRIES, enabling farmers to plow and plant their spring wheat. A Chinook originates as AIR descends on the eastern or LEEWARD side of the Rocky

Mountains. Having lost all its moisture on the WINDWARD side, this is a dry wind that warms ADIABATICALLY as it descends. The wind is welcomed by farmers and is sometimes called the "snow-eater." In Europe, similar winds are called FÖHN.

Chlorofluorocarbons (CFCs). Manufactured compounds, not occurring in nature, consisting of chlorine, fluorine, and carbon. CFCs are stable and have heat-absorbing properties, so they have been used extensively for cooling in refrigeration and air-conditioning units. Previously, they were used as propellants for aerosol products. CFCs rise into the STRATOSPHERE where ULTRAVIOLET RADIATION causes them to react with OZONE, changing it to oxygen and exposing the earth to higher levels of ultraviolet (UV) radiation. Therefore, the manufacture and use of CFCs was banned in many countries. The commercial name for CFCs is Freon.

Chorology. Description or mapping of a REGION. Also known as chorography.

Chronometer. Highly accurate CLOCK or timekeeping device. The first accurate and effective chronometers were constructed in the mid-eighteenth century by John Harrison, who realized that accurate time-keeping was the secret to NAVIGATION at SEA.

Chubasco. Type of severe STORM that occasionally occurs in the Gulf of California and along the west COAST of Mexico.

Cinder cone. Small conical HILL produced by PYROCLASTIC materials from a VOLCANO. The material of the cone is loose SCORIA.

Volcanic cinder cones on the island of Hawaii. (Corbis)

Circle of illumination. Line separating the sunlit part of the earth from the part in darkness. The circle of illumination moves around the earth once in every approximately 24 hours. At the VERNAL and autumnal EQUINOXES, the circle of illumination passes through the POLES.

Washington State's Wenatchee Mountains contain remnants of old alpine glaciers, which surround Mount Stuart; these include U-shaped valleys and four small cirque glaciers in the shadows. (U.S. Geological Survey)

Cirque. Circular BASIN at the head of an ALPINE GLACIER, shaped like an armchair. Many cirques can be seen in MOUNTAIN areas where glaciers have completely melted since the last ICE AGE.

Cirro. Prefix meaning high CLOUDS, from the Latin word *cirrus,* meaning a lock of hair.

Cirrocumulus. High, thin, puffy white CLOUDS of ice crystals that look like ripples. They appear between 20,000 and 40,000 feet (6,000-12,000 meters) above the earth's surface. One type of cirrocumulus cloud is called a "mackerel sky," because the clouds resemble large fish scales, especially when they are colored pink at SUNSET.

Cirrostratus. Semitransparent sheets of CLOUD, comprising layers of thin ice crystals. They appear between 20,000 and 40,000 feet (6,000-12,000 meters) above the earth's surface. A halo around the MOON can be caused by cirrostratus clouds.

Cirrus. High, wispy tufts of CLOUDS, white but almost transparent because they are composed mostly of ice crystals. Formed at a height of 20,000 to 30,000 feet (6,000-9,000 meters). Cirrus clouds can indicate an approaching COLD FRONT. Sometimes called "mares' tails." The prefix *cirro* is added to shape words to define two other kinds of high clouds—CIRROCUMULUS and CIRROSTRATUS.

City Beautiful movement. Planning and architectural movement that was at its height from around 1890 to the 1920's in the United States. It was believed that classical architecture, wide and carefully laid-out streets, parks, and urban monuments would reflect the higher values of the society and be a civilizing, even uplifting, experience for the citizens of such cities. Civic pride was fostered through remodeling or modernizing older URBAN AREAS. Chicago, Illinois, and Pasadena, California, are cities where the planners of the City Beautiful movement left their imprint.

City. Generally large human SETTLEMENTS in which nonagricultural occupations dominate. In the United States, a city is technically defined as an incorporated MUNICIPALITY with definite boundaries and legal powers set forth in a charter granted by the STATE. Since 1910 the U.S. Census Bureau has recognized any PLACE with more than twenty-five hundred inhabitants as URBAN. In the United Kingdom, cities were historically defined not by their POPULATION sizes, but on the basis of their religious status: whether they had cathedrals with bishops.

Civilization. Type of CULTURE or society comprising urban POPULATIONS, RELIGION, architecture, and formalized methods of passing on learning. The AGRICULTURAL REVOLUTION, when human societies began growing crops instead of relying on HUNTING AND GATHERING, enabled the earliest civilizations to emerge more than six thousand years ago, in Mesopotamia.

Clastic. ROCK or sedimentary matter formed from fragments of older rocks.

Clay. Finely grained SOIL. Soils that are largely clay are generally unsuitable for AGRICULTURE because they are impermeable to plant roots. Soils known as cracking clays, or vertisols, can absorb large amounts of water, which causes them to swell as they expand, but cracks as large as three feet (1 meter) deep open as the soil dries out. These clays are found in Texas and over large REGIONS in eastern Australia, India, and tropical East Africa.

Clearing. Open part of a FOREST where trees and other VEGETATION have been removed, often for farming.

Cliff. Hillslope that is nearly vertical.

Climagraph. See CLIMOGRAPH.

Climate. Long-term conditions of TEMPERATURE and PRECIPITATION for a PLACE over a period of not less than thirty years. Climate takes account of variability and extremes to give a composite picture of a climate type or a climate REGION. Climate is not exactly the same as WEATHER, which is the situation of the ATMOSPHERE at any moment, and thus changes constantly.

Climatology. Study of Earth CLIMATES by analysis of long-term WEATHER patterns over a minimum of thirty years of statistical records. Climatologists—scientists who study climate—seek similarities to enable grouping into climatic REGIONS. Climate patterns are closely related to natural VEGETATION. Computer TECHNOLOGY has enabled investigation of phenomena such as the EL NIÑO effect and global climate change. The KOEPPEN CLIMATE CLASSIFICATION system is the most commonly used scheme for climate classification.

Climograph. Graph that plots TEMPERATURE and PRECIPITATION for a selected LOCATION. The most commonly used climographs plot monthly temperatures and monthly precipitation, as used in the KOEPPEN CLIMATE CLASSIFICATION. Also spelled "climagraph." The term climagram is rarely used.

Clinometer. Instrument used by surveyors to measure the ELEVATION of land or the inclination (slope) of the land surface.

Clock. Machine that measures time and displays the result continuously. An especially accurate clock or timekeeper is called a CHRONOMETER.

Clockwise. Rotating direction matching that of the hands on a CLOCK dial when viewed from the same perspective. This term and its opposite, COUNTERCLOCKWISE, are often used to describe the movements of WEATHER phenomena and the ROTATIONS of celestial objects.

Cloud. Atmospheric occurrence of moisture droplets and ice crystals suspended in AIR. Particles such as DUST or smoke may also be present. Clouds are classified according to their shapes and heights. The classification, and the words used, were proposed by English scientist Luke Howard in the early nineteenth century. See ALTOCUMULUS; ALTOSTRATUS; CIRROCUMULUS; CIRROSTRATUS; CIRRUS; CUMULONIMBUS; CUMULUS.

Cloud cover. Amount of the sky that is covered with CLOUD, shown on a WEATHER MAP by shading parts of a circle. If the sky is half-covered, the right half of the circle is shaded.

Cloud cover is the amount of sky covered with clouds. On weather maps, it is shown by shading parts of a circle. When the sky is half-covered, the right half of the circle is shaded. (PhotoDisc)

Cloud-free. Having less than 30 percent CLOUD COVER, allowing clear imaging of a surface area.

Cloud seeding. Injection of CLOUD-nucleating particles into likely clouds to enhance PRECIPITATION.

Cloudburst. Heavy rain that falls suddenly.

Coal. One of the FOSSIL FUELS. Coal was formed from fossilized plant material, which was originally FOREST. It was then buried and compacted, which led to chemical changes. Most coal was formed during the CARBONIFEROUS PERIOD (286 million to 360 million years ago) when the earth's CLIMATE was wetter and warmer than at present.

Coast. Land above the high-tide level where land meets the OCEAN. The coast is a PLACE of constant change, due to natural changes, such as varying SEA LEVEL or TECTONIC movements, as well as human activities, such as constructing PORT facilities, marinas, and housing developments.

Coastal plain. Large area of flat land near the OCEAN. Coastal plains can form in various ways, but FLUVIAL DEPOSITION is an important process. In the United States, the coastal plain extends from Texas to North Carolina.

Coastal wetlands. Shallow, wet, or flooded shelves that extend back from the freshwater-saltwater interface and may consist of MARSHES, BAYS, LAGOONS, tidal flats, or MANGROVE SWAMPS.

Coastline. Specific line of contact between land and SEA. The coastline changes constantly because of TIDES, STORMS, and sea-level changes. Also called SHORELINE.

Cognitive map. Mental image that each person has of the world, which includes LOCATIONS and connections. These maps expand as children mature, from plans of their rooms, to their houses, to their neighborhoods. Adults know certain parts of the CITY and the streets connecting them. See also MENTAL MAP.

Coke. Type of fuel produced by heating COAL.

Col. Lower section of a RIDGE, usually formed by the headward EROSION of two CIRQUE GLACIERS at an ARÊTE. Sometimes called a saddle.

Cold cloud. Visible SUSPENSION of tiny ice crystals, supercooled water droplets, or both at sub-freezing TEMPERATURES.

Cold front. FRONT or leading edge of an advancing cold AIR mass that displaces warmer air as it moves. On a WEATHER MAP, a cold front is shown by a line of triangular "shark teeth" pointing in the direction of advance. A cold front is accompanied by STORMS and rain.

Cold War. Period that lasted from the end of World War II, in 1945, until the collapse of the Soviet Union, in 1991, during which the communist NATIONS of the East and the noncommunist nations of the West competed for world supremacy and engaged in military buildups in anticipation of a new global war.

Cold wave. Sudden onset of extremely cold WEATHER, with TEMPERATURE below freezing, the change taking less than twenty-four hours.

Colonial cities. Cities established and developed by colonial governments to serve as administrative or commercial centers. Some colonial cities were newly created in a LOCATION where there was previously no URBAN SETTLEMENT. The colonial power laid out a new planned CITY with ceremonial buildings and PLACES, offices, administrative headquarters, commercial facilities, and military barracks. Local people came to the new colonial city to serve in low-paid service jobs such as clerks and servants. Colonial cities of this kind include Calcutta, Nairobi, Hong Kong, and Jakarta. A different type of colonial city arose through the addition of colonial functions to an already established settlement. There, source of labor and considerable SITE advantages already existed. Mexico City, Delhi, and Shanghai are examples of this kind of colonial city. Most colonial cities were located on the COAST, for ease of access to shipping goods back to the European colonial power.

Colonialism. Control of one COUNTRY over another STATE and its people. Many European countries have created colonial empires, including Great Britain, France, Spain, Portugal, the Netherlands, and Russia.

Colony. COUNTRY that is a political DEPENDENCY of another NATION. During the early twentieth century, most of the countries of Africa, the Pacific, and the Caribbean, as well as many in Asia, were colonies of European powers. By the beginning of the twenty-first century, however, few colonies remained in the world.

Columbian exchange. Interaction that occurred between the Americas and Europe after the voyages of Christopher Columbus. Food crops from the New World transformed the diet of many European countries.

Combe. Welsh word for the uppermost part of a VALLEY, above the springline. Also called coombe.

Comet. Small body in the SOLAR SYSTEM, consisting of a solid head with a long gaseous tail. The elliptical ORBIT of a comet causes it to range

The most famous of the many comets that pass through the Solar System is Halley's, which made its last transit near Earth in 1986. (PhotoDisc)

from very close to the SUN to very far away. In ancient times, the appearance of a comet in the sky was thought to be an omen of great events or changes, such as war or the death of a king.

Comfort index. Number that expresses the combined effects of TEMPERATURE and HUMIDITY on human bodily comfort. The index number is obtained by measuring ambient conditions and comparing these to a chart.

Commodity chain. Network linking labor, production, delivery, and sale for any product. The chain begins with the production of the raw material, such as the extraction of MINERALS by miners, and extends to the acquisition of the finished product by a consumer.

Communications. Systems used to transmit messages or information from one PLACE to another; now systems such as the Internet, telephones, television, and mail.

Communities, animal or plant. See BIOME.

Compass, magnetic. Instrument that determines direction, used for NAVIGATION. A magnetic needle is mounted so that it can rotate and align its ends with the earth's MAGNETIC FIELD. A naturally occurring ORE of iron called lodestone aligns in a north-south direction; a piece of iron, placed in contact with the lodestone, becomes magnetized and also aligns itself this way. The magnetic compass has been used since the twelfth century, both in Europe and in China. The earliest compasses consisted of a magnetized needle that floated in a bowl of water. Soon a card with the points of the compass was added to the compass, so that readings could be made quickly and simply. When ships were built of iron in the nineteenth century, many adaptations had to be made to maintain the accuracy of the magnetic compass.

Complex crater. IMPACT CRATER of large diameter and low depth-to-diameter ratio caused by the presence of a central UPLIFT or ring structure.

Composite cone. Cone or VOLCANO formed by volcanic explosions in which the LAVA is of different composition, sometimes fluid, sometimes PYROCLASTS such as cinders. The alternation of layers allows a concave shape for the cone. These are generally regarded as the world's most beautiful volcanoes. Composite volcanoes are sometimes called STRATOVOLCANOES.

Condensation. Process in which water changes from a vapor state to a liquid state, releasing heat into the surrounding AIR; this process is the opposite of EVAPORATION, which requires the input of heat. Water VAPOR condenses into DEW, FOG, or CLOUD droplets.

Condensation nuclei. Microscopic particles that may have originated as DUST, soot, ASH from fires or VOLCANOES, or even SEA SALT; an essential part of CLOUD formation. When AIR rises and cools to the DEW POINT (saturation), the moisture droplets condense around the nuclei, leading to the creation of raindrops or snowflakes. A typical air mass might contain ten billion condensation nuclei in a single cubic yard (1 cubic meter) of air.

Cone, volcanic. See Cinder cone; Composite cone.

Cone of depression. Cone-shaped depression produced in the water table by pumping from a well.

Confined aquifer. Aquifer that is completely filled with water and whose upper boundary is a confining bed; it is also called an artesian aquifer.

Confining bed. Impermeable layer in the earth that inhibits vertical water movement.

Confluence. Place where two streams or rivers flow together and join. The smaller of the two streams is called a tributary.

Conglomerate. Type of sedimentary rock consisting of smaller rounded fragments naturally cemented together by another mineral. If the cemented fragments are jagged or angular, the rock is called breccia.

Conglomerate corporation. Large, transnational corporation whose operations cover a diverse range of economic activities. Generally created by mergers and acquisitions; one of the effects of globalization of industry and trade. United States' names dominate the conglomerates: Phillip Morris expanded from tobacco into foods, beer, real estate, and publishing. Nestlé, based in Switzerland, sells many food products, pet food, wine, and cosmetics, and also has interests in the hotel business.

Conical projection. Map projection that can be imagined as a cone of paper resting like a witch's hat on a globe with a light source at its center; the images of the continents would be projected onto the paper. In reality, maps are constructed mathematically. A conic projection can show only part of one hemisphere. This projection is suitable for constructing a map of the United States, as a good equal-area representation can be achieved. Also called conic projection.

Coniferous forest. Forest type found naturally growing in cool climates with sufficient precipitation, throughout most of Canada and

Coniferous forests are found naturally growing in cool climates with at least moderate precipitation. (PhotoDisc)

extensive areas of Russia, where it is called TAIGA. The trees are needleleaf species of pine, fir, spruce, and larch. BOREAL FOREST is another name for this BIOME. The trees are valuable sources of lumber for construction or pulpwood for newspaper production. Needleleaf forests also occur in mountainous REGIONS, as in parts of the Rocky Mountains, Sierra Nevada, European Alps, and Himalayas.

Consequent river. RIVER that flows across a LANDSCAPE because of GRAVITY. Its direction is determined by the original slope of the land. TRIBUTARY streams, which develop later as EROSION proceeds, are called subsequent streams.

Conservationism. Practice of protecting natural things from loss or damage—from a stand of trees to the earth's ENVIRONMENT as a whole. SOIL conservation deals with preventing the loss of valuable TOPSOIL through poor agricultural practices. Open space conservation leads to the creation of parkland reserves around and sometimes within cities. Many international and other organizations are involved in conservation of RESOURCES such as RAIN FORESTS, plant and animal species, and natural areas.

Contaminant. Any ion or chemical that is introduced into the ENVIRONMENT, especially in concentrations greater than those normally present.

Continent. Principal LANDMASSES of the earth, comprising Eurasia, Africa, North America, South America, Antarctica, and Australia. The continents cover approximately one-quarter of the earth's surface at present sea-level conditions, accounting for an area of almost 60 million square miles (150 million sq. km.). The continental shelves, which are now under water, are geologically part of the continents; if this total area were measured, the continents would account for about one-third of the earth's surface. These continents are based on a physical definition, but many people use a cultural distinction, and thus divide Eurasia into two continents—Asia and Europe—because there are marked cultural differences between peoples of the two parts of this single landmass. Throughout geologic time, the continents have been joined and separated many times. See also PLATE TECTONICS.

Continental climate. CLIMATE experienced over the central REGIONS of large LANDMASSES; drier and subject to greater seasonal extremes of TEMPERATURE than at the CONTINENTAL MARGINS.

Continental crust. Earth's CRUST consists of two different types of ROCKS: continental crust and OCEANIC CRUST. Continental crust is crystalline, and lighter in weight than the denser oceanic crust. GRANITE is the most abundant rock of the continental crust. An older term for continental crust was "SIAL."

Continental divide. High REGION that separates DRAINAGE on a continental scale. In North America, the Continental Divide separates RIVERS flowing west to the Pacific Ocean from those flowing south and east to the Atlantic Ocean or north to the Arctic Ocean.

Continental drift. Theory, proposed in 1912 by German scientist Alfred Wegener, holding that the CONTINENTS of the earth have changed position continuously over the past 225 million years. Wegener hypothesized the existence of an ancient SUPERCONTINENT, which he named PANGAEA, that slowly broke apart as its component continents drifted into their present positions. Although the theory was not initially accepted in English-speaking countries, it became the influential theory of PLATE TECTONICS.

Continental glaciers. Continental glaciers once covered much of northern North America and Europe, but now only Greenland and Antarctica have such huge masses of permanent ice and SNOW. Because continental glaciers are so large, they affect the CLIMATE of large REGIONS outside their boundaries by lowering AIR and water TEMPERATURES. Continental glaciers of past ICE AGES have produced a wide variety of erosional and depositional features in northern LATITUDES.

Continental island. ISLAND that is part of the CONTINENTAL SHELF, rising above SEA LEVEL. Such islands are actually part of their adjacent CONTINENTS, with the same compositions as their nearby continents. They have become separated from the continents as a result of TECTONIC movement over thousands of years. Their ROCK types are consistent with the mainland. Most of the world's large islands are continental islands. For example, Greenland is a continental island of North America, New Guinea is a continental island of Australia, and Madagascar is a continental island of Africa. In contrast, OCEANIC ISLANDS, such as Hawaii, rise from the OCEAN floor. Some continental islands are located close to the COAST and have become separated only since sea level rose over the last several thousand years. Tasmania is an example of this kind of continental island. These continental islands are called high islands.

Continental margin. COASTLINE and BEACHES, plus the CONTINENTAL SHELF and continental rise. Continental margins make up about 20 percent of the OCEAN but are the most valuable part because of land values onshore and fishing in the shallow waters just offshore. According to the theory of PLATE TECTONICS, continental margins can be passive, when there is no obvious movement of the margin, although it is being moved with the PLATE; at an active continental margin, in contrast, motion is obvious. The continental margin of California, Oregon, and Washington is an active continental margin, where TRANSFORM MOTION is occurring with SUBDUCTION occurring to the north of this.

Continental rift zones. Continental rift zones are PLACES where the CONTINENTAL CRUST is stretched and thinned. Distinctive features include active VOLCANOES and long, straight VALLEY systems formed by normal FAULTS. Continental rifting in some cases has evolved into the breaking apart of a CONTINENT by SEAFLOOR SPREADING to form a new OCEAN.

Continental shelf. Shallow, gently sloping part of the seafloor adjacent to the mainland. The continental shelf is geologically part of the CONTI-NENT and is made of CONTINENTAL CRUST, whereas the OCEAN floor is OCEANIC CRUST. Although continental shelves vary greatly in width, on average they are about 45 miles (75 km.) wide and have slopes of 7 minutes (about one-tenth of a DEGREE). The average depth of a continental shelf is about 200 feet (60 meters). The outer edge of the continental shelf is marked by a sharp change in angle where the CONTI-NENTAL SLOPE begins. Most continental shelves were exposed above current SEA LEVEL during the PLEISTOCENE EPOCH and have been submerged by rising sea levels over the past eighteen thousand years.

Continental shield. Area of a CONTINENT that contains the oldest ROCKS on Earth, called CRATONS. These are areas of granitic rocks, part of the CONTINENTAL CRUST, where there are ancient MOUNTAINS. The Canadian Shield in North America is an example.

Continental slope. Part of the OCEAN floor between the outer edge of the CONTINENTAL SHELF and the DEEP seafloor of the OCEAN BASINS. Although there is great variation, the continental slope on average is about 12 miles (20 km.) in width and has an average slope of 4 DE-GREES; it extends from a water depth of about 425 feet (130 meters) below SEA LEVEL to somewhere between 5,000 and 10,000 feet (1,400-3,000 meters) deep.

Contour lines. Lines on a TOPOGRAPHIC MAP that join PLACES of equal EL-EVATION. A series of contour lines reveals the overall shape and elevation of TERRAIN.

Convection. Transfer of heat from a source area to a point farther away through vertical motion and subsequent spreading, as in the vertical AIR circulation in which warm air rises and cool air sinks.

Convectional rain. Type of PRECIPITATION caused when AIR over a warm surface is warmed and rises, leading to ADIABATIC cooling, CONDENSA-TION, and, if the air is moist enough, rain.

Convective overturn. Renewal of the bottom waters caused by the sinking of SURFACE WATERS that have become denser, usually because of decreased TEMPERATURE.

Convergence (climate). AIR flowing in toward a central point.

Convergence (physiography). Process that occurs during the second half of a SUPERCONTINENT CYCLE, whereby crustal PLATES collide and intervening OCEANS disappear as a result of plate SUBDUCTION.

Convergent plate boundary. Compressional PLATE BOUNDARY at which an oceanic PLATE is subducted or two continental plates collide.

Convergent plate margin. Area where the earth's LITHOSPHERE is returned to the MANTLE at a SUBDUCTION ZONE, forming volcanic "IS-LAND ARCS" and associated HYDROTHERMAL activity.

Conveyor belt current. Large CYCLE of water movement that carries warm water from the north Pacific westward across the Indian Ocean, around Southern Africa, and into the Atlantic, where it warms the ATMOsphere, then returns at a deeper OCEAN level to rise and begin the process again.

Coombe. See COMBE.

Coordinated universal time (UTC). International basis of time, introduced to the world in 1964. The basis for UTC is a small number of ATOMIC CLOCKS. Leap seconds are occasionally added to UTC to keep it synchronized with universal time.

Copse. English term for a small area where the VEGETATION consists of small trees and thick shrubs or BUSHES. An older word is "coppice."

Coral reef. LIMESTONE structure found in shallow tropical SEAS, consisting of a living biological community atop the calcium carbonate remains of many generations of dead coral. Individual coral polyps are tiny, but their accreted skeletons can form huge RIDGES or REEFS. Depending on LOCATION, reefs are classified into four types—FRINGING REEFS, barrier reefs, ATOLLS, and patch reefs. The world's largest coral reef is the Great Barrier Reef, off the east COAST of northern Australia. Reefs also are found in Florida and around many ISLANDS of the Pacific and Indian Oceans.

The Florida Keys comprise small sandy islands built up by wave action on coral reefs. (Visit Florida)

Cordillera. Large mountain CHAIN such as the Rocky Mountains in North America or the Andes Mountains in South America. "Cordilleran system" denotes this group of relatively young mountains, extending from Alaska in the north to Tierra del Fuego in the south.

Core. Innermost part of the earth, believed to comprise two distinct zones. The OUTER CORE is dense, molten, and mostly iron, and is responsible for the earth's MAGNETIC FIELD; the INNER CORE is thought to be solid and mostly iron.

Core-mantle boundary. SEISMIC discontinuity 1,790 miles (2,890 km.) below the earth's surface that separates the MANTLE from the OUTER CORE.

Core region. Area, generally around a COUNTRY'S CAPITAL CITY, that has a large, dense POPULATION and is the center of TRADE, financial services, and production. The rest of the country is referred to as the PERIPHERY. On a larger scale, the CONTINENT of Europe has a core region, which includes London, Paris, and Berlin; Iceland, Portugal, and Greece are peripheral LOCATIONS.

Coriolis effect. Apparent deflection of moving objects above the earth because of the earth's ROTATION. The deflection is to the right in the NORTHERN HEMISPHERE and to the left in the SOUTHERN HEMISPHERE. The deflection is inversely proportional to the speed of the earth's rotation, being negligible at the EQUATOR but at its maximum near the POLES. The Coriolis effect is a major influence on the direction of surface WINDS. Sometimes called Coriolis force.

Corn Belt. Part of the United States covering Iowa, Illinois, and Indiana, and parts of Minnesota, South Dakota, Nebraska, Minnesota, and Ohio. A REGION of mixed crop-and-LIVESTOCK AGRICULTURE, where corn growing and hog farming are combined on farms. Corn is also used at feedlots to fatten cattle from areas farther west. The main agricultural product of the Corn Belt, therefore, is meat.

Corrasion. EROSION and lowering of a STREAMBED by FLUVIAL action, especially by ABRASION of the bedload (material transported by the STREAM) but also including SOLUTION by the water.

Cosmogony. Study of the origin and nature of the SOLAR SYSTEM.

Cosmopolitanism. Intellectual openness to a variety of CULTURES, experiences, and products from other REGIONS or countries. Previously, people gained a cosmopolitan perspective mainly through travel to foreign countries, but now television, motion pictures, and the Internet can bring aspects of foreign contemporary cultures directly to consumers in their own homes.

Cotton Belt. Part of the United States extending from South Carolina through Georgia, Alabama, Mississippi, Tennessee, Louisiana, Arkansas, Texas, and Oklahoma, where cotton was grown on PLANTATIONS using slave labor before the Civil War. After that war, the South stagnated for almost a century. Racial SEGREGATION contributed to cultural isolation from the rest of the United States. Cotton is still produced in this REGION, but California has overtaken the Southern STATES as a cotton producer, and other agricultural products, such as soybeans and poultry, have become dominant crops in the old Cotton

Belt. In-migration, due to the SUN BELT attraction, has led to rapid urban growth in the old Cotton Belt..

Counterclockwise. Rotating direction opposite to that of the hands on a
CLOCK dial when viewed from the same perspective. This term and its
opposite, CLOCKWISE, are often used to describe the movements of
WEATHER phenomena and the ROTATIONS of celestial objects. For example, low-pressure areas are characterized by WINDS moving in a
counterclockwise direction in the NORTHERN HEMISPHERE.

Counterurbanization. Out-migration of people from URBAN AREAS to
smaller TOWNS or RURAL areas. As large modern cities are perceived to
be overcrowded, stressful, polluted, and dangerous, many of their residents move to areas they regard as more favorable. Such moves are often related to individuals' retirements; however, younger workers and
families are also part of counterurbanization.

Country. Commonly used to mean an independent and sovereign STATE,
such as the United States, Canada, or Germany; a NATION-STATE. Also
used to mean RURAL, as compared with a TOWN or CITY, as in "country
roads" or "country cousins."

County. Unit into which some countries are subdivided for local administration. In the United States, the level below the STATE government
(called PARISHES in Louisiana). In the United Kingdom, the level of major division for administration, similar to the states of the United States.

County seat. CITY or TOWN containing the administrative headquarters
of the surrounding COUNTY.

Cove. Small opening in the COASTLINE of any larger body of water. A cove
can be a small BAY, usually well protected by HEADLANDS.

Crag. Scottish and Welsh word for a steep, rocky CLIFF in the MOUNTAINS
or on coastal HEADLANDS and ISLANDS.

Crater. Circular depression at the top of a VOLCANO, from which molten material emerges. Craters also are found on the flanks of larger volcanoes.

*Wizard Island in Oregon's Crater Lake. The lake fills the caldera of a volcano that erupted about
76,000 years ago. After the volcano emptied itself, it collapsed into its own hole to create the
caldera. (Corbis)*

Crater morphology. Structure or form of CRATERS and the related processes that developed them.

Craton. Large, geologically old, relatively stable CORE of a continental
LITHOSPHERIC PLATE, sometimes termed a CONTINENTAL SHIELD.

Creep. Slow, gradual downslope movement of SOIL materials under gravitational stress. Creep tests are experiments conducted to assess the effects of time on ROCK properties, in which environmental conditions
(surrounding pressure, TEMPERATURE) and the deforming stress are
held constant.

Crestal plane. Plane or surface that goes through the highest points of all
beds in a fold; it is coincident with the axial plane when the axial plane
is vertical.

Cretaceous era. Third, last, and longest geologic PERIOD of the MESOZOIC
ERA, 144 million to 65 million years ago. During the era SEAS covered
much of North America and the Rocky Mountains were formed. The
end of the era was marked by the EXTINCTION of the dinosaurs.

*Inside a crevasse in Blue
Ice Valley on the Greenland
ice sheet, sixty-five feet
below the surface. (U.S.
Geological Survey)*

Crevasse. Deep vertical crack that forms in a GLACIER as a result of
stresses. Fresh snowfall can cover a crevasse, making glacier exploration hazardous.

Crop rotation. Agricultural practice of growing alternating crops on the
same field. Generally a LEGUME, such as alfalfa or clover, is grown as a

fodder crop after a grain crop has been grown for one or two years. The legume helps restore soil fertility by adding nitrogen. Crop rotation was developed in the late seventeenth century in Europe, as one of a series of advances known as the AGRICULTURAL REVOLUTION.

Cross-bedding. Layers of ROCK or SAND that lie at an angle to horizontal bedding or to the ground.

Crown land. Land belonging to a NATION's MONARCHY. Some parts of crown land are used as public parks; others are leased and used for private agriculture or other commercial purposes.

Crude birth rate. Ratio of the number of live births in a COUNTRY in a single year for every thousand people of the total POPULATION. In high-income economies, the crude birth rate is less than twenty. In the United States in 2000, it was fifteen per thousand. In some African countries, such as Somalia, it was fifty per thousand.

Crude death rate. Ratio of the number of deaths in a COUNTRY in a single year for every thousand people of the total POPULATION. When the CRUDE BIRTH RATE is higher than the crude death rate, the population of a country is increasing assuming that there is no net MIGRATION loss. In the United States in 2000, the crude death rate was nine per thousand. The difference between the crude birth rate and the crude death rate is the rate of NATURAL INCREASE, which is expressed as a percentage. For the United States in 2000, the rate of natural increase was 0.6 percent (six per thousand).

Crude oil. Unrefined OIL, as it occurs naturally. Also called PETROLEUM.

Crust. Outer layer of the earth, made of crystalline ROCKS and varying in thickness from 3 miles (5 km.) beneath the OCEANS to 38 miles (60 km.) under the continental mountain RANGES. It consists of rocky material which is less dense than the MANTLE.

Crustal movements. PLATE TECTONICS theorizes that Earth's CRUST is not a single rigid shell, but comprises a number of large pieces that are in motion, separating or colliding. There are two types of crust—the older continental and the much younger OCEANIC CRUST. When PLATES diverge, at SEAFLOOR SPREADING zones, new (oceanic) crust is created from the MAGMA that flows out at the MID-OCEAN RIDGES. When plates converge and collide, denser oceanic crust is SUBDUCTED under the lighter CONTINENTAL CRUST. The boundaries at the areas where plates slide laterally, neither diverging nor converging, are called TRANSFORM FAULTS. The San Andreas Fault represents the world's best-known transform BOUNDARY. As a result of crustal movements, the earth can be deformed in several ways. Where PLATE BOUNDARIES converge, compression can occur, leading to FOLDING and the creation of SYNCLINES and ANTICLINES. Other stresses of the crust can lead to fracture, or faulting, and accompanying EARTHQUAKES. LANDFORMS created in this way include HORSTS, GRABEN, and BLOCK MOUNTAINS.

Cuesta. Spanish term used to describe an ESCARPMENT and its associated gentle dip slope, formed in SEDIMENTARY ROCKS.

Cultural ecology. Study of the interaction between humans and their ENVIRONMENT; for example, the study of how human societies have adapted to certain physical conditions, such as prolonged cold in northern Canada or arid conditions in northern Mexico. Cultural ecology is related to cultural anthropology. In highly urbanized societies, cultural ecologists study how people shape their URBAN environments and how those environments affect human lifestyle and behavior.

Cultural geography. Study of how LANDSCAPES, space, and PLACE shape various CULTURES, while at the same time different cultures shape and influence the landscapes, spaces, and places.

Cultural landscape. Evidence of the effect of human activities on the natural LANDSCAPE.

Cultural nationalism. Movement that has grown rapidly in the face of globalization. Modern media have a homogenizing effect on national and regional CULTURES. Governments, and some smaller groups, have attempted to protect and preserve their culture from globalization, which has often meant Americanization. Some measures adopted have included negative acts such as restricting the broadcast of American programs on television and banning certain recordings of videocassettes; positive measures include promotion of literature in the national LANGUAGE or regional DIALECT, investment in local cultural and artistic productions, and the creation of archives, oral histories, and museums.

Culture. In anthropology, the learned parts of human behavior that are transmitted through generations. Culture includes LANGUAGE, RELIGION, foods and their preparation, clothing, ceremonies, housing, and the other factors that are shared by a cultural group. Geographers are concerned with how any particular culture, or way of life, is related to the physical LANDSCAPE. They also study how cultures vary from one PLACE to another and how cultures change over time.

Culture hearth. LOCATION in which a CULTURE has developed; a CORE REGION from which the culture later spread or diffused outward through a larger REGION. Mesopotamia, the Nile Valley, and the Peruvian ALTIPLANO are examples of culture hearths.

Cumulonimbus. Huge, dense CLOUDS that can rise up into the STRATOSPHERE. Cumulonimbus clouds produce LIGHTNING and THUNDERSTORMS, so the base of a cumulonimbus cloud can be dark while the top is gleaming white. The flat or anvil-shaped top of the cloud is sometimes called a thunderhead. See also CLOUDS.

Cumulus. Puffy CLOUDS ranging from small to extremely large. Cumulus clouds occur below 6,500 feet (2,000 meters). These clouds are sometimes compared to cotton balls or cauliflower. Coastal REGIONS see cumulus clouds every DAY.

Curie point. TEMPERATURE at which a magnetic MINERAL locks in its magnetization. Also known as Curie temperature.

Cycle. Sequence of naturally recurring events and processes. Most cycles consume ENERGY to move a substance through the ENVIRONMENT.

Cycle of erosion. Influential MODEL of LANDSCAPE change proposed by William Morris Davis near the end of the nineteenth century. The UPLIFT of a relatively flat surface, or PLAIN, in an area of moderate RAINFALL and TEMPERATURE, led to gradual EROSION of the initial surface in a sequence Davis categorized as Youth, Maturity, and Old Age. The final landscape was called PENEPLAIN. Davis also recognized the stage of REJUVENATION, when a new uplift could give new ENERGY to the cycle, leading to further downcutting and erosion. The model also was used to explain the sequence of LANDFORMS developed in REGIONS of ALPINE GLACIERS. The model has been criticized as misleading, since CRUSTAL MOVEMENT is continuous and more frequent than Davis perhaps envisaged, but it remained useful as a description of TOPOGRAPHY. Also known as the Davisian cycle or geomorphic cycle.

Cyclone. Low-pressure system of rotating WINDS, converging and ascending. In the NORTHERN HEMISPHERE, the rotation is COUNTERCLOCKWISE; in the SOUTHERN HEMISPHERE, the rotation is CLOCKWISE. See also ANTICYCLONE; HURRICANE; TROPICAL CYCLONE; TYPHOON.

Cyclonic rain. In the NORTHERN HEMISPHERE winter, two low-pressure systems or CYCLONES—the Aleutian Low and the Icelandic Low—develop over the OCEAN near 60 DEGREES north LATITUDE. The polar FRONT forms where the cold and relatively dry ARCTIC AIR meets the warmer, moist air carried by westerly WINDS. The warm air is forced upward, cools, and condenses. These cyclonic STORMS often move south, bringing winter PRECIPITATION to North America, especially to the STATES of Washington and Oregon.

Cylindrical projection. MAP PROJECTION that represents the earth's surface as a rectangle. It can be imagined as a cylinder of paper wrapped around a globe with a light source at its center; the images of the CONTINENTS would be projected onto the paper. In reality, MAPS are constructed mathematically. It is impossible to show the North Pole or South Pole on a cylindrical projection. Although the map is conformal, distortion of area is extreme beyond 50 DEGREES north and south LATITUDES. The Mercator projection, developed in the sixteenth century by the Flemish cartographer Gerhardus Mercator, is the best-known cylindrical projection. It has been popular with seamen because the shortest route between two PORTS (the GREAT CIRCLE route) can be plotted as straight lines that show the COMPASS direction that should be followed. Use of this projection for other purposes, however, can lead to misunderstandings about size; for example, compare Greenland on a globe and on a Mercator map. See also MAPS.

Dale. English word for a VALLEY.

Dam. Structure built across a RIVER to control the flow of water. It is thought that the earliest dams were constructed to store water for IRRIGATION during the dry part of the year. Modern dams store water for cities and industry and also produce hydroelectricity. The LAKE that forms artificially behind a dam is called a RESERVOIR. Every dam must have a SPILLWAY so excess water is released when the level in the reservoir becomes too high. As engineering TECHNOLOGY led to the construction of huge dams in the twentieth century, many critics voiced opposition. The waters of the reservoir often inundated areas of great scenic beauty, valuable agricultural land, and historic structures. Large dams also displaced many people.

Dams are structures built across streams to control the flow of water. Lakes that form artificially behind dams are called reservoirs. (PhotoDisc)

Date line. See INTERNATIONAL DATE LINE.

Datum level. Baseline or level from which other heights are measured, above or below. MEAN SEA LEVEL is the datum commonly used in surveying and in the construction of TOPOGRAPHIC MAPS.

Davisian cycle. See CYCLE OF EROSION.

Dawn. Period of time from the first appearance of sunlight in the morning to when the SUN is fully above the HORIZON. The length of time varies with LATITUDE, being shortest at the EQUATOR. At the POLES, dawn lasts for about seven weeks during the summer months.

Day. Interval of time between successive passages of the SUN or star over a MERIDIAN of the earth.

Daylight saving time. System of seasonal adjustments in CLOCK settings designed to increase hours of evening sunlight during summer months.

In the spring, clocks are set ahead one hour; in the fall, they are put back to standard time. In North America, these changes are made on the first Sunday in April and the last Sunday in October. The U.S. Congress standardized daylight saving time in 1966; however, parts of Arizona, Indiana, and Hawaii do not follow the system.

Death rate. Annual number of deaths per one thousand individuals of a given POPULATION. For the United States, the death rate was nine persons (per thousand per year) at the end of the twentieth century. Shortened form of "CRUDE DEATH RATE."

Débâcle. In a scientific context, this French word means the sudden breaking up of ice in a RIVER in the spring, which can lead to serious, sudden flooding.

Debris avalanche. Large mass of SOIL and ROCK that falls and then slides on a cushion of AIR downhill rapidly as a unit.

Debris flow. Flowing mass consisting of water and a high concentration of SEDIMENT with a wide RANGE of size, from fine muds to coarse gravels.

Deciduous forest. Mixed, broadleaf FOREST that was once common in moist, temperate CLIMATES in the United States, Europe, and Asia.

Birch trees can be found in boreal forests throughout the world. (PhotoDisc)

Common trees were oak, ash, elm, walnut, maple, and birch. The leaves of the trees turn red or yellow as the WEATHER becomes cool, and the branches are bare throughout the winter when TEMPERA-TURES fall below freezing. Centuries of clearing have destroyed significant portions of deciduous forests.

Declination, magnetic. Measure of the difference, in DEGREES, between the earth's NORTH MAGNETIC pole and the North Pole on a MAP; this difference changes slightly each year. The needle of a magnetic COM-PASS points to the earth's geomagnetic pole, which is not exactly the same as the North Pole of the geographic GRID or the set of lines of LATITUDE and LONGITUDE. The geomagnetic poles, north and south, mark the ends of the AXIS of the earth's MAGNETIC FIELD, but this field is not stationary. In fact, the geomagnetic poles have completely reversed hundreds of times throughout earth history. Lines of equal magnetic declination are called ISOGONIC LINES.

Declination of the Sun. LATITUDE of the SUBSOLAR POINT, the PLACE on the earth's surface where the SUN is directly overhead. In the course of a year, the declination of the Sun migrates from 23.5 DEGREES north LATI-TUDE, at the (northern) summer SOLSTICE, to 23.5 degrees south latitude, at the (northern) WINTER SOLSTICE. Hawaii is the only part of the United States that experiences the Sun directly overhead twice a year.

Deep. Relatively deep part of an OCEAN, part of the ABYSSAL PLAIN.

Deep ecology. View or philosophy of nature that has two major aspects. Self-realization is the view that humans are merely one part of a complex world system with many different parts. Egalitarianism, or biospherical egalitarianism, places the whole earth at the center of life and holds that every species has the same rights; humans are not superior to, or more important than, any other species, or even ROCKS. Deep ecologists argue that humans should respect the nonhuman world and not regard it as merely a means to sustain human life.

Deep-focus earthquakes. EARTHQUAKES occurring at depths ranging from 40 to 400 miles (65 to 650 km.) below the earth's surface. This RANGE of depths represents the zone from the base of the earth's CRUST to approximately one-quarter of the distance into Earth's MAN-TLE. Deep-focus earthquakes provide scientists information about the PLANET's interior structure, its composition, and SEISMICITY. Observation of deep-focus earthquakes has played a fundamental role in the discovery and understanding of PLATE TECTONICS.

Deep-ocean currents. Deep-ocean currents involve significant vertical and horizontal movements of seawater. They distribute oxygen- and nutrient-rich waters throughout the world's OCEANS, thereby enhancing biological productivity.

Deep-sea plain. See ABYSSAL PLAIN.

Defile. Narrow MOUNTAIN PASS or GORGE through which troops could march only in single file.

Example of deflation—a block of granite hollowed out by windblown sand in Chile's Atacama Province. (U.S. Geological Survey)

Deflation. EROSION by WIND, resulting in the removal of fine particles. The LANDFORM that typically results is a deflation hollow.

Deforestation. Removal or destruction of FORESTS. In the late twentieth century, there was widespread concern about tropical deforestation— destruction of the tropical RAIN FOREST—especially that of Brazil. Forest clearing in the TROPICS is uneconomic because of low SOIL fertility.

Clear-cutting of forests for commercial timber is one of the major contributors to deforestation. (PhotoDisc)

Deforestation causes severe EROSION and environmental damage; it also destroys habitat, which leads to the EXTINCTION of both plant and animal species.

Degradation. Process of CRATER EROSION from all processes, including WIND and other meteorological mechanisms. See also DENUDATION.

Degree (geography). Unit of LATITUDE or LONGITUDE in the geographic GRID, used to determine ABSOLUTE LOCATION. One degree of latitude is about 69 miles (111 km.) on the earth's surface. It is not exactly the same everywhere, because the earth is not a perfect sphere. One degree of longitude varies greatly in length, because the MERIDIANS converge at the POLES. At the EQUATOR, it is 69 miles (111 km.), but at the North or South Pole it is zero.

Degree (temperature). Unit of MEASUREMENT of TEMPERATURE, based on the CELSIUS SCALE, except in the United States, which uses the FAHRENHEIT SCALE. On the Celsius scale, one degree is one-hundredth of the difference between the freezing point of water and the boiling point of water.

Dehydration. Release of water from pore spaces or from hydrous MINERALS as a result of increasing TEMPERATURE.

Delta. Area of DEPOSITION of ALLUVIUM where a RIVER enters the SEA or a LAKE. The Greek letter "delta" was used to describe the MOUTH of the Nile River; not all river deltas have this shape. Some are elongated along DISTRIBUTARIES and are called bird's-foot deltas, for example the Mississippi River. The largest delta in the world is the combined delta of the Ganges and Brahmaputra Rivers in Bangladesh.

The combined delta of the Ganges and Brahmaputra Rivers in Bangladesh is the world's largest. (PhotoDisc)

Demographic measure. Statistical data relating to POPULATION.

Demographic transition. MODEL of POPULATION change that fits the experience of many European countries, showing changes in birth and death rates. In the first stage, in preindustrial countries, population size was stable because both BIRTH RATES and DEATH RATES were high. Agricultural reforms, together with the INDUSTRIAL REVOLUTION and subsequent medical advances, led to a rapid fall in the death rate, so that the second and third stages of the model were periods of rapid population growth, often called the POPULATION EXPLOSION. In the fourth stage of the model, birth rates fall markedly, leading again to stable population size.

Demography. Study of POPULATION, especially of changes measured by such statistics as BIRTH RATES, DEATH RATES, and MIGRATION.

Dendritic drainage. Most common pattern of STREAMS and their TRIBUTARIES, occurring in areas of uniform ROCK type and regular slope. A MAP, or aerial photograph, shows a pattern like the veins on a leaf—smaller streams join the main stream at an acute angle.

Denudation. General word for all LANDFORM processes that lead to a lowering of the LANDSCAPE, including WEATHERING, mass movement, EROSION, and transport.

Dependency. Territory, such as a COLONY or PROTECTORATE, ruled by a NATION of which it is not an integral part.

Deposition. Laying down of SEDIMENTS that have been transported by water, WIND, or ice.

Depression. Term used in European countries for a midlatitude CYCLONE or low-pressure system. PRECIPITATION usually results as these systems move from west to east across Europe or North America.

Deranged drainage. LANDSCAPE whose integrated drainage network has been destroyed by irregular glacial DEPOSITION, yielding numerous shallow LAKE BASINS.

Derivative maps. MAPS that are prepared or derived by combining information from several other maps.

Desalinization. Process of removing SALT and MINERALS from seawater or from saline water occurring in AQUIFERS beneath the land surface to render it fit for AGRICULTURE or other human use.

Desert. Large REGION of dry CLIMATE, which consequently has a sparse human POPULATION. The desert BIOME occupies about one-quarter of the earth's lands and comprises a distinctive assemblage of FLORA and FAUNA with specific adaptations to this physical ENVIRONMENT. Desert plants are mostly small and sparse and have XEROPHYTIC characteristics. Animals are small and often nocturnal. The hot deserts of the world are located in northern Africa through southwest Asia, and in Australia, the southwest United States, Chile, and southwest Africa. There also are cool or temperate deserts, located in the northern United States and extensively in central Asia, where ELEVATION is part of the reason

The spectacular sand dunes of Asia's Gobi Desert display the constantly changing ridges and shapes caused by wind. (Digital Stock)

for this difference. Desert LANDFORMS are quite distinctive. Although SAND DUNES are popularly associated with deserts, they cover only 10 percent of the world's deserts. Stony, mountainous desert LANDSCAPES are much more common, especially in United States deserts.

Desert climate. Low PRECIPITATION, low HUMIDITY, high daytime TEMPERATURES, and abundant sunlight are characteristics of desert climates. The hot DESERTS of the world generally are located on the western sides of CONTINENTS, at LATITUDES from fifteen to thirty DEGREES north or south of the EQUATOR. One definition, based on precipitation, defines deserts as areas that receive between 0 and 9 inches (0 to 250 millimeters) of precipitation per year. REGIONS receiving more precipitation are considered to have a SEMIDESERT climate, in which some AGRICULTURE is possible.

Desert pavement. Surface covered with smoothed PEBBLES and gravels, found in arid areas where DEFLATION (WIND EROSION) has removed smaller particles. Called a "gibber plain" in Australia and a *reg* in Arabic-speaking countries. See also ERG.

Desertification. Increase in DESERT areas worldwide, largely as a result of overgrazing or poor agricultural practices in semiarid and marginal CLIMATES. DEFORESTATION, DROUGHT, and POPULATION increase also contribute to desertification. The REGION of Africa just south of the Sahara Desert, known as the SAHEL, is the largest and most dramatic demonstration of desertification.

Detrital minerals. See DETRITUS.

Detrital rock. SEDIMENTARY ROCK composed mainly of grains of silicate MINERALS as opposed to grains of calcite or CLAYS.

Detritus. MINERALS which have been eroded, transported, and deposited as SEDIMENTS. Also called detrital minerals.

Development. Level of INDUSTRIALIZATION and standard of living in a COUNTRY. Economic geographers study various measures of development. The countries of the world can be divided into four levels of economic development: high-income, upper-middle-income, lower-middle-income, and low-income. The low-income countries are concentrated in Africa and Asia. In the past, terms such as "undeveloped," "less developed," and "underdeveloped" were used for the various low-income to lower-middle-income economies, as was the now-outdated term THIRD WORLD.

Devolution. Breaking up of a large COUNTRY into smaller independent political units is the final and most extreme form of devolution. The Soviet Union devolved from one single country into fifteen separate countries in 1991. At an intermediate level, devolution refers to the granting of political autonomy or self-government to a REGION, without a complete split. The reopening of the Scottish Parliament in 1999 and the Northern Ireland parliament in 2000 are examples of devolution; the Parliament of the United Kingdom had previously met only in London and made laws there for all parts of the country. Canada experienced devolution with the creation of the new territory of Nunavut, whose residents elect the members of their own legislative assembly.

Dew. Deposit of water droplets on objects whose surface has sufficiently cooled, generally by loss of heat through nighttime RADIATION, to a TEMPERATURE sufficiently low to condense water vapor from the surrounding AIR. See also HOAR FROST.

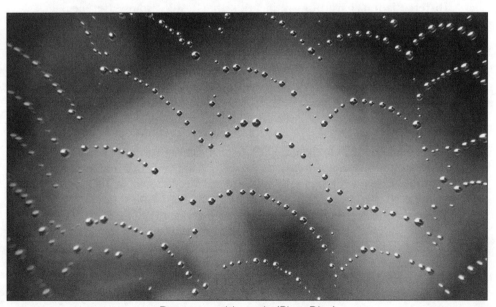

Dew on a spider web. (PhotoDisc)

Dew point. TEMPERATURE at which an AIR mass becomes saturated and can hold no more moisture. Further cooling leads to CONDENSATION. At ground level, this produces DEW.

Diagenesis. Conversion of unconsolidated SEDIMENT into consolidated ROCK after burial by the processes of compaction, cementation, recrystallization, and replacement.

Dialect. Regional variation of a standard LANGUAGE. It can consist of different pronunciations and different word usage. Speakers of the language can understand the dialect, even if they do not speak it. Within the United States, there are several regional dialects; for example, people in Texas speak a different dialect from people in Boston.

Diaspora. Dispersion of a group of people from one CULTURE to a variety of other REGIONS or to other lands. A Greek word, used originally to refer to the Jewish diaspora. Jewish people now live in many countries, although they have Israel as a HOMELAND. Similar to this are the diasporas of the Irish and the Chinese.

Diastrophism. Deformation of the earth's CRUST by faulting or FOLDING.

Diatom ooze. Deposit of soft mud on the OCEAN floor consisting of the shells of diatoms, which are microscopic single-celled creatures with SILICA-rich shells. Diatom ooze deposits are located in the southern Pacific around Antarctica and in the northern Pacific. Other PELAGIC, or deep-ocean, SEDIMENTS include CLAYS and calcareous ooze.

Differential weathering. Physical and CHEMICAL WEATHERING that occurs at irregular or different rates, caused by variations in composition and resistance of a ROCK or by differences in intensity of WEATHERING, and usually resulting in an uneven surface where more resistant material stands higher or protrudes above less resistant parts.

Differentiation. Layering within ROCK that results from differences in density; the lighter material rises to the surface while the heaviest material sinks to the bottom of a mixture of substances.

Diffusion. Process of growth and spread outward from a center or core area over time. It is applied to many phenomena in CULTURAL GEOGRAPHY, such as the spread of disease, or the growth of a CITY, LANGUAGE, and ideas. Modern telecommunications make possible almost instantaneous diffusion of ideas, images, and sounds throughout the developed world.

Dike (geology). LANDFORM created by IGNEOUS intrusion when MAGMA or molten material within the earth forces its way in a narrow band through overlying ROCK. The dike can be exposed at the surface through EROSION.

Dike (water). Earth wall or DAM built to prevent flooding; an EMBANKMENT or artificial LEVEE. Sometimes specifically associated with structures built in the Netherlands to prevent the entry of seawater. The land behind the dikes was reclaimed for AGRICULTURE; these new fields are called POLDERS.

Dingle. Old English word for a small, secluded VALLEY with trees.

Distance-decay function. Rate at which an activity diminishes with increasing distance. The effect that distance has as a deterrent on human activity is sometimes described as the FRICTION OF DISTANCE. It occurs because of the time and cost of overcoming distances between people and their desired activity. An example of the distance-decay function is the rate of visitors to a football stadium. The farther people have to travel, the less likely they are to make this journey.

Distributary. STREAM that takes waters away from the main CHANNEL of a RIVER. A DELTA usually comprises many distributaries. Also called distributary channel.

Distribution. Way in which some feature, or group of features, under examination is spread out over a REGION. Geographers look for patterns of distribution and seek explanations for the patterns.

Diurnal range. Difference between the highest and lowest TEMPERATURES registered in one twenty-four-hour period.

Diurnal tide. Having only one high tide and one low tide each lunar DAY; TIDES on some parts of the Gulf of Mexico are diurnal.

Divergence. Process of fracturing and dissecting a SUPERCONTINENT, thereby creating new oceanic ROCK; divergence represents the initial half of the supercontinent CYCLE.

Divergent boundary. BOUNDARY that results where two TECTONIC PLATES are moving apart from each other, as is the case along MID-OCEANIC RIDGES.

Divergent margin. Area where the earth's CRUST and LITHOSPHERE form by SEAFLOOR SPREADING.

Divergent plates. TECTONIC PLATE BOUNDARY where two PLATES are moving apart.

Diversity. Variety of life, usually described in terms of the number of species present.

Divide. RIDGE that separates one DRAINAGE BASIN from the adjoining basin. The CONTINENTAL DIVIDES of the United States separate those STREAMS that flow to the Pacific Ocean, the Gulf of Mexico, the Atlantic Ocean, Hudson Bay, and the Arctic Ocean. Also known as drainage divide.

Doctor. WINDS or BREEZES that bring relief from unpleasant or oppressive WEATHER conditions. In Western Australia, the cool sea breeze that comes in the afternoon is called the Fremantle Doctor. A similar phenomenon in South Africa is called the Cape Doctor. The names come from an earlier time, but these winds have become important in reducing AIR POLLUTION in the cities of Perth and Cape Town, respectively.

Doldrums. Narrow BELT of OCEANS on both sides of the THERMAL EQUATOR (INTERTROPICAL CONVERGENCE ZONE) which is a zone of calms or light variable WINDS. In the days of travel by sailing ship, sailors feared becoming trapped in this part of the world. The epic poem *The Rime of* *the Ancient Mariner* (1857) by Samuel Taylor Coleridge describes the imaginary plight of a vessel caught in the doldrums. Now, the term "in the doldrums" has come to mean a feeling of downheartedness.

Doline. Large SINKHOLE or circular depression formed in LIMESTONE areas through the CHEMICAL WEATHERING process of carbonation.

Dolomite. MINERAL consisting of calcium and magnesium carbonate compounds that often forms from PRECIPITATION from seawater; it is abundant in ancient ROCKS.

Dome. Small circular structure formed by FOLDING or warping of the earth's CRUST, such as are found among the Ozark Mountains and the Black Hills of South Dakota.

Domestication. Change from wild to tame or suitable for human agricultural use. The domestication of animals and plants, which began several thousand years ago, led to the farming of crops and the grazing of animals, making CIVILIZATION possible.

Donga. South African word for a dry STREAMBED in an arid area. See also ARROYO, WADI, and WASH.

Double cropping. In warm moist CLIMATES, farmers can produce two crops from the same field in a single year. This is the case with rice growing in southeast Asian countries and in southern China. Fertile SOILS are an advantage in these REGIONS as well.

Downburst. Downward outflowing of AIR and the associated WIND shear from a THUNDERSTORM that is especially hazardous to aircraft.

Downland. Flat to rolling UPLAND area or PLATEAU, covered with grass and used mainly for grazing sheep. Often known as downs.

Downwelling. Sinking of OCEAN water.

Drainage. Collection and removal of water from PRECIPITATION by STREAM CHANNELS. The geology of an area influences the drainage pattern. Typically, a DENDRITIC pattern forms in an area of uniform slope and rocktype. A CENTRIPETAL drainage pattern indicates a GRABEN. On a DOME or VOLCANO, a radial pattern forms. In areas of alternating hard and soft ROCKS, usually areas of FOLDING, a trellis pattern is seen. See also INTERNAL DRAINAGE.

Drainage basin. Area of the earth's surface that is drained by a STREAM. Drainage basins vary greatly in size, but each is separated from the next by RIDGES, or drainage DIVIDES. The CATCHMENT of the drainage basin is the WATERSHED.

Drainage density. Total length of all STREAMS in a DRAINAGE BASIN divided by the area of that basin. A humid CLIMATE has a high drainage density, while DESERTS have a low drainage density.

Drainage divide. See DIVIDE.

Drift ice. ARCTIC or ANTARCTIC ice floating in the open SEA.

Drizzle. Very fine rain, comprising small raindrops.

Drought. Prolonged period with no PRECIPITATION; abnormally dry WEATHER sufficiently prolonged for the lack of precipitation to cause a serious HYDROLOGICAL imbalance. Drought is a relative rather than an absolute condition, but the end result is a water shortage for an activity such as plant growth or for some group of people such as farmers.

Drowned valley. Feature occurring where a SHORELINE has been submerged, usually through rising SEA LEVEL. Where a series of long HEADLANDS and alternating ESTUARIES occur, this is called a RIA COAST. The northeastern coast of the United States has many long narrow INLETS that were RIVER VALLEYS when sea level was lower.

Drumlin. Low HILL, shaped like half an egg, formed by DEPOSITION by CONTINENTAL GLACIERS. A drumlin is composed of TILL, or mixed-size materials. The wider end faces upstream of the glacier's movement; the tapered end points in the direction of the ice movement. Drumlins usually occur in groups or swarms.

Dune. Deposits of SAND of various shapes—crescents, RIDGES, and heaps. SAND DUNES are moved by WINDS.

Duricrust. See LATERITE.

Dust. Particles smaller than 62 micrometers in size. Dust particles are moved great distances by WINDS, from one CONTINENT to another. Dust particles may come from SOIL EROSION, as was the case in the famous DUST BOWL of the 1930's.

Dust Bowl. Part of the southwestern Great PLAINS of the United States, in Kansas, Texas, Oklahoma, New Mexico, and Colorado. In the 1920's, GRASSLAND that formerly had been used for cattle grazing was plowed up for grain growing. This fragile ENVIRONMENT then suffered from poor agricultural practices, combined with a long DROUGHT in the 1930's. The TOPSOIL was carried away as DUST by strong WINDS. Daylight was obscured by clouds of dust so thick that streetlights burned throughout the DAY in cities such as Kansas City and St. Louis. During the Great Depression, thousands of people from the Dust Bowl abandoned their farms, many moving to California in hopes of a better life there. The federal government acted to stabilize the REGION in the 1940's.

Dust devil. Whirling cloud of DUST and small debris, formed when a small patch of the earth's surface becomes heated, causing hot AIR to rise;

Dust devils. (PhotoDisc)

cooler air then flows in and begins to spin. The resulting dust devil can grow to heights of 150 feet (50 meters) and reach speeds of 35 miles (60 km.) per hour. See WILLY WILLY.

Dust dome. Dome of AIR POLLUTION, composed of industrial gases and particles, covering every large CITY in the world. The pollution sometimes is carried downwind to outlying areas.

Dust storm. Particles such as DUST transported long distances by WINDS. The size of the particles influences the distance traveled. STORMS from the Sahara Desert can carry dust even north of the Alps in Europe. Removal of particles by the wind is called DEFLATION.

Early Paleozoic. That part of geologic history that is somewhat younger than about 550 million years before the present.

Earth pillar. Formation produced when a boulder or caprock prevents EROSION of the material directly beneath it, usually CLAY. The clay is easily eroded away by water during RAINFALL, except where the overlying ROCK protects it. The result is a tall, slender column, as high as 20 feet (6.5 meters) in exceptional cases.

Earth radiation. Portion of the electromagnetic spectrum, from about 4 to 80 microns, in which the earth emits about 99 percent of its RADIATION.

Earth tide. Slight deformation of Earth resulting from the same forces that cause OCEAN TIDES, those that are exerted by the MOON and the SUN.

Earthflow. Term applied to both the process and the LANDFORM characterized by fluid downslope movement of SOIL and ROCK over a discrete plane of failure; the landform has a HUMMOCKY surface and usually terminates in discrete lobes.

Earthquake. Movement of the earth's CRUST when there is a sudden release of built-up ENERGY along a FAULT. WAVES travel through the crust, as well as through the underlying MANTLE. Earthquake intensity is measured using the moment magnitude scale. Before 1993, the RICHTER SCALE was commonly used. An early descriptive scale of earthquake intensity is the MERCALLI SCALE.

Earthquake focus. Area below the surface of the earth where active movement occurs to produce an EARTHQUAKE.

Earthquake swarm. Number of EARTHQUAKES that occur close together and closely spaced in time.

Earthquake waves. Vibrations that emanate from an EARTHQUAKE; earthquake waves can be measured with a SEISMOGRAPH.

Earth's core. See CORE.

Earth's heat budget. Balance between the incoming SOLAR RADIATION and the outgoing terrestrial reradiation.

Eastern Hemisphere. The half of the earth containing Europe, Asia, and Africa; generally understood to fall between LONGITUDES 20 DEGREES west and 160 degrees east.

Ebb tide. Outgoing or falling TIDE that, in most parts of the world, occurs twice in a 24-hour period. See also FLOOD TIDE.

Eclipse. Event where all or part of the light emitted, or reflected, by an astronomical object is obscured by another astronomical object.

Eclipse, lunar. Obscuring of all or part of the light of the MOON by the shadow of the earth. A lunar eclipse occurs at the full moon up to three times a year. The surface of the Moon changes from gray to a reddish color, then back to gray. The sequence may last several hours.

Eclipse, solar. At least twice a year, the SUN, MOON, and Earth are aligned in one straight line. At that time, the Moon obscures all the light of the Sun along a narrow band of the earth's surface, causing a total eclipse; in REGIONS of Earth adjoining that area, there is a partial eclipse. A corona (halo of light) can be seen around the Sun at the total eclipse. Viewing a solar eclipse with naked eyes is extremely dangerous and can cause blindness.

Solar eclipse observed from North Dakota on February 26, 1979. (PhotoDisc)

Ecliptic. Intersection of the plane of the earth's ORBIT with the celestial sphere; with the exception of Pluto, the orbital planes of the other PLANETS lie within 7 DEGREES of the ecliptic.

Ecliptic, plane of. Imaginary plane that would touch all points in the earth's ORBIT as it moves around the SUN. The angle between the plane of the ecliptic and the earth's AXIS is 66.5 DEGREES.

Ecological imperialism. Introduction of foreign and exotic species of plants and animals into ECOSYSTEMS. The interchange between the Old World and the New World was often deliberate and beneficial (see COLUMBIAN EXCHANGE), but there were also unintentional introductions. These often had severely detrimental effects on native species and led to problems with feral animals and similar pests like starlings and pigeons, or to noxious weeds such as thistles and dandelions.

Ecology. Science that studies the relationship between living organisms (plants and animals) and their ENVIRONMENT. Ecologists also study individual ECOSYSTEMS in detail.

Economy. System of production, DISTRIBUTION, and consumption of goods and services, usually within a single COUNTRY. Measures of the strength of a country's economy include the GROSS DOMESTIC PRODUCT per capita or the gross national product per capita. The growth of transnational enterprises and international TRADE has led to a global economy.

Ecosystem. Association of living and nonliving parts of a group of plants and animals and their physical ENVIRONMENT.

Edaphic. Related to the SOIL. Edaphic factors that influence plants include HUMUS content, which is related to fertility; soil TEXTURE; soil structure; and the presence of various soil organisms such as bacteria and earthworms.

Eddy. Mass of water that is spun off an OCEAN CURRENT by the current's meandering motion.

Edge cities. Forms of suburban downtown in which there are nodal concentrations of office space and shopping facilities. Edge cities are located close to major freeways or highway intersections, on the outer edges of METROPOLITAN AREAS.

Effective temperature. TEMPERATURE of a PLANET based solely on the amount of SOLAR RADIATION that the planet's surface receives; the effective temperature of a planet does not include the GREENHOUSE temperature enhancement effect.

Ejecta. Material ejected from the CRATER made by a meteoric impact.

Ekman layer. REGION of the SEA, from the surface to about 100 meters down, in which the WIND directly affects water movement.

Ekman spiral. Water movement in lower depths of an ocean that occurs at a slower rate and in a different direction from SURFACE WATER movement.

El Niño. Conditions—also known as El Niño-Southern Oscillation (ENSO) events—that occur every two to ten years and affect WEATHER and OCEAN TEMPERATURES, particularly off the COAST of Ecuador and Peru. Most of the time, the Peru, or Humboldt, Current causes cold, nutrient-rich water to well up off the coast of Ecuador and Peru. During ENSO years, the cold UPWELLING is replaced by warmer SURFACE WATER that does not support PLANKTON and fish. Fisheries decline and seabirds starve. Climatic changes of El Niño can bring FLOODS to normally dry areas and DROUGHT to wet areas. Effects can extend across North and South America, and to the western Pacific Ocean. During the 1990's, the ENSO event fluctuated but did not vanish completely, which caused tremendous damage to fisheries and AGRICULTURE, STORMS and droughts in North America, and numerous HURRICANES.

Elevation. Vertical distance of a point on the earth's surface above or below MEAN SEA LEVEL.

Ellipse. Shape of Earth's ORBIT; rather than a circle with one center, the ellipse has two foci with the SUN located at one of the foci.

Eluviation. Removal of materials from the upper layers of a SOIL by water. Fine material may be removed by SUSPENSION in the water; other material is removed by SOLUTION. The removal by solution is called LEACH-ING. Eluviation from an upper layer leads to illuviation in a lower layer.

Embankment. Artificial earthen mound built to support a road or to control the movement of water.

Emigration. Leaving one's COUNTRY of birth to settle permanently in another country. See also IMMIGRATION.

Emirate. Islamic NATION ruled by a monarch whose title is emir.

Enclave. Piece of territory completely surrounded by another COUNTRY. Two examples are Lesotho, which is surrounded by the Republic of South Africa, and the Nagorno-Karabakh REGION, populated by Armenians but surrounded by Azerbaijan. The term is also used for smaller regions, such as ethnic neighborhoods within larger cities. See also EXCLAVE.

Endemic. Found in a particular PLACE and no other.

Endemic species. Species confined to a restricted area in a restricted ENVIRONMENT.

Endogenic sediment. SEDIMENT produced within the water column of the body in which it is deposited; for example, calcite precipitated in a LAKE in summer.

Energy. Scientifically, the capacity to do work. Geographers study sources of energy such as FOSSIL FUELS, which are a NONRENEWABLE RESOURCE; ALTERNATIVE (RENEWABLE) ENERGY forms, such as SOLAR energy, HYDROELECTRIC POWER, TIDAL ENERGY, WIND power, and GEOTHERMAL energy; and NUCLEAR ENERGY, which is an abundant RESOURCE but presents serious problems with the disposal of radioactive waste.

ENSO. Acronym for EL NIÑO-Southern Oscillation, used to denote the complete linked atmospheric/OCEAN phenomenon.

Environment. Surroundings of an organism, or a group of organisms, which enable it to survive. Several physical factors are involved in the creation of a suitable natural environment, including TEMPERATURE, moisture, food supply, and waste removal or recycling. The natural environment is sometimes modified extensively by humans, through cooling and heating of buildings, for example. Humans are also concerned with the social environment and the cultural environment.

Environmental degradation. Situation that occurs in slum areas and SQUATTER SETTLEMENTS because of poverty and inadequate INFRASTRUCTURE. Too-rapid human POPULATION growth can lead to the accumulation of human waste and garbage, the POLLUTION of GROUNDWATER, and DENUDATION of nearby FORESTS. As a result, LIFE EXPECTANCY in such degraded areas is lower than in the RURAL communities from which many of the settlers came. INFANT MORTALITY is particularly high. When people leave an area because of such environmental degradation, that is referred to as ecomigration.

Environmental determinism. Theory that the major influence on human behavior is the physical ENVIRONMENT. Some evidence suggests that

TEMPERATURE, PRECIPITATION, sunlight, and TOPOGRAPHY influence human activities. Originally espoused by early German geographers, this theory has led to some extreme stances, however, by authors who have sought to explain the dominance of Europeans as a result of a cool temperate CLIMATE.

Environmental ethics. Philosophy or view of nature that believes humans should always apply moral principles to their treatment of nature and natural phenomena. In other words, the moral values and judgments that are applied to relations between humans should enable people to decide what is right and good with respect to nature. In its extreme form, environmental ethics could hold that ROCKS are of equal value to humans, or that insects such as mosquitoes should have the same rights as humans to a safe and happy life. See also DEEP ECOLOGY.

Environmental justice. Belief that it is unfair to locate many factories, dumps, and hazardous waste facilities in low-socioeconomic areas of cities, or on a global scale, in countries with low-income economies. Advocates of environmental justice emphasize that economic inequality is inevitable in capitalist society, but that it is immoral and even illegal to pollute neighborhoods on the basis of low economic status.

Eocene epoch. Part of the CENOZOIC ERA, dating to about 37 million years ago.

Eolian (aeolian). Relating to, or caused by, WIND. In Greek mythology, Aeolus was the ruler of the winds. EROSION, TRANSPORT, and DEPOSITION are common eolian processes that produce LANDFORMS in DESERT REGIONS.

Eolian deposits. Material transported by the WIND and later deposited.

Eolian erosion. Mechanism of EROSION or CRATER degradation caused by WIND.

Eon. Largest subdivision of geologic time; the two main eons are the PRECAMBRIAN (c. 4.6 billion years ago to 544 million years ago) and the PHANEROZOIC (c. 544 million years ago to the present).

Epeiric sea. Shallow SEA that temporarily (in geologic terms) covers a portion of a CRATON; also termed an EPICONTINENTAL SEA.

Ephemeral stream. Watercourse that has water for only a DAY or so.

Epicenter. Spot on the earth's surface directly above the focus of an EARTHQUAKE. Shock waves produced by the earthquake radiate outward from the focus, allowing SEISMOLOGISTS to locate the epicenter quickly. The RICHTER SCALE, which was used to calculate earthquake magnitude until the 1990's, measured the amplitude of SEISMIC WAVES recorded at least 60 miles (100 km.) from the epicenter. Sometimes a previously unknown FAULT is revealed, a blind fault that does not appear as a surface break. The Northridge earthquake in California in 1994 revealed the LOCATION of a blind THRUST FAULT, where movement had occurred along the FAULT LINE directly beneath the epicenter at Northridge.

Epicontinental sea. Shallow SEAS that are located on the CONTINENTAL SHELF, such as the North Sea or Hudson Bay. Also called an EPEIRIC SEA.

Epifauna. Organisms that live on the seafloor.

Epilimnion. Warmer surface layer of water that occurs in a LAKE during summer stratification; during spring, warmer water rises from great depths, and it heats up through the summer SEASON.

Epoch. Unit of geologic time; a subdivision of a PERIOD.

Equal-area projection. MAP PROJECTION that maintains the correct area of surfaces on a MAP, although shape distortion occurs. The property of such a map is called equivalence. See also MAPS.

Equator. Imaginary line of LATITUDE around the earth's circumference at its widest part, lying equidistant from the POLES and perpendicular to the earth's axis. The equator is a GREAT CIRCLE that divides the earth into two equal halves, the NORTHERN and SOUTHERN HEMISPHERES.

Equinox. Period of equal DAY and night, twelve hours of each, everywhere on Earth, occurring when the CIRCLE OF ILLUMINATION passes through both the POLES. The VERNAL (spring) equinox falls on March 21 and the autumnal (fall) equinox on September 22 in the NORTHERN HEMI-SPHERE; the SEASONS are reversed in the SOUTHERN HEMISPHERE.

Era. One of the major divisions of geologic time, including one or more PERIODS.

Erg. Sandy DESERT, sometimes called a SEA of SAND. Erg deserts account for less than 30 percent of the world's deserts. "Erg" is an Arabic word.

Erosion. Wearing down and carrying away of earth surface materials by water, WIND, ice, or WAVES. See also CYCLE OF EROSION.

Erratic. See GLACIAL ERRATIC.

Eruption, volcanic. Emergence of MAGMA (molten material) at the earth's surface as LAVA. There are various types of volcanic eruptions, depending on the chemistry of the magma and its viscosity. Scientists refer to effusive and explosive eruptions. Low-viscosity magma generally produces effusive eruptions, where the lava emerges gently, as in Hawaii and Iceland, although explosive events can occur at those SITES as well. Gently sloping

The spectacular cliffs of Parador, Spain, are the product of eons of erosion. (PhotoDisc)

SHIELD VOLCANOES are formed by effusive eruptions; FLOODS, such as the Columbian Plateau, can also result. Explosive eruptions are generally associated with SUBDUCTION. Much gas, including steam, is associated with magma formed from OCEANIC CRUST, and the compressed gas helps propel the explosion. COMPOSITE CONES, such as Mount Saint Helens, are created by explosive eruptions.

Escarpment. Steep slope, often almost vertical, formed by faulting. Sometimes called a FAULT SCARP.

Esker. Deposit of coarse gravels that has a sinuous, winding shape. An esker is formed by a STREAM of MELTWATER that flowed through a tunnel it formed under a CONTINENTAL GLACIER. Now that the continental glaciers have melted, eskers can be found exposed at the surface in many PLACES in North America.

Estuarine zone. Area near the COASTLINE that consists of estuaries and coastal saltwater WETLANDS.

Estuary. PLACE where the MOUTH OF A RIVER enters the SEA, causing FRESH WATER and SALT WATER to mix. Tidal EBB and flow occur in an estuary. Estuaries are WETLANDS that are productive ECOSYSTEMS.

Etesian winds. WINDS that blow from the north over the Mediterranean during July and August.

Ethnic group. Group of people with a distinctive CULTURE, usually including RELIGION, LANGUAGE, traditions, and customs, and sometimes racial ancestry.

Ethnic religion. RELIGION associated with a particular ETHNIC GROUP that does not actively seek to convert others to the same religious beliefs. Judaism and Hinduism are good examples of ethnic religions. Religions that actively seek converts are called proselytic religions; Christianity and ISLAM are examples.

Ethnocentrism. Belief that one's own ETHNIC GROUP and its CULTURE are superior to any other group.

Ethnography. Study of different CULTURES and human societies.

Eustacy. Any change in global SEA LEVEL resulting from a change in the absolute volume of available sea water. Also known as eustatic sea-level change.

Eustatic movement. Changes in SEA LEVEL.

Evaporation. Change from liquid water to water vapor as water molecules enter the ATMOSPHERE. The process is the opposite of CONDENSATION.

Evapotranspiration. Combined word for EVAPORATION and TRANSPIRATION. Both processes transfer moisture from the earth's surface to the ATMOSPHERE in the form of water vapor: evaporation from water in the OCEANS and other water bodies; transpiration from VEGETATION.

Exclave. Territory that is part of one COUNTRY but separated from the main part of that country by another country. Alaska is an exclave of the United States; Kaliningrad is an exclave of Russia. See also ENCLAVE.

Yosemite National Park's Half Dome is perhaps the world's most famous example of an exfoliation dome. (PhotoDisc)

Exfoliation. When GRANITE rocks cooled and solidified, removal of the overlying rock that was present reduced the pressure on the granite mass, allowing it to expand and causing sheets or layers of rock to break off. An exfoliation DOME, such as Half Dome in Yosemite National Park, is the resultant LANDFORM.

Exosphere. REGION beyond the earth's ATMOSPHERE, 300 miles (500 km.) above the earth's surface. Only a few atoms of hydrogen and helium are thought to exist in the exosphere.

Exotic stream. RIVER that has its source in an area of high RAINFALL and then flows through an arid REGION or DESERT. The Nile River is the most famous exotic STREAM. In the United States, the Colorado River is a good example of an exotic stream.

Expansion-contraction cycles. Processes of wetting-drying, heating-cooling, or freezing-thawing, which affect SOIL particles differently according to their size.

Expansive soils. Expansive soils, SOILS that expand and contract with the gain and loss of water, cause billions of dollars in damage to houses, other lightweight structures, and pavements, exceeding the costs incurred by EARTHQUAKES and flooding.

External economies. Cost savings that firms can enjoy by choosing a LOCATION close to functionally related or similar activities. For example, a soft-drink plant would be better off to locate close to a glass-making factory that would supply bottles with a low cost of transport.

Extinction. Disappearance of a species or large group of animals or plants.

Extrusive rock. Fine-grained, or glassy, ROCK which was formed from a MAGMA that cooled on the surface of the earth.

Eye. Calm central REGION of a HURRICANE, composed of a tunnel with strong sides.

The eye of Hurricane Elena can be easily seen in this September, 1985, photography taken from the space shuttle Discovery. (Corbis)

Fahrenheit scale. TEMPERATURE scale with the freezing point of water at 32 DEGREES (0 degrees Celsius) and its boiling point at 212 degrees (100 degrees Celsius). In the year 2000, the United States was the only major COUNTRY using the Fahrenheit scale instead of the CELSIUS SCALE.

Fall line. Edge of an area of uplifted land, marked by WATERFALLS where STREAMS flow over the edge.

Famine. Severe shortage of food, often caused by DROUGHT.

Fata morgana. Large mirage. Originally, the name given to a multiple mirage phenomenon often observed over the Straits of Messina and supposed to be the work of the fairy ("fata") Morgana. Another famous fata morgana may be seen in Antarctica.

Fathom. MEASUREMENT of water depth used by mariners. A fathom is 6 feet (1.83 meters).

Fathometer. Instrument that uses sound waves or sonar to determine the depth of water or the depth of an object below the water.

Fault. Fracture of the earth's CRUST, usually as a result of an EARTHQUAKE.

Fault-block mountain. See BLOCK MOUNTAIN.

Fault drag. Bending of ROCKS adjacent to a FAULT.

Fault line. Line of breakage on the earth's surface. FAULTS may be quite short, but many are extremely long, even hundreds of miles. The origin of the faulting may lie at a considerable depth below the surface. Movement along the fault line generates EARTHQUAKES.

Fault line in a plowed field.

Fault plane. Angle of a FAULT. When fault blocks move on either side of a fault or fracture, the movement can be vertical, steeply inclined, or sometimes horizontal. In a NORMAL FAULT, the fault plane is steep to almost vertical. In a REVERSE FAULT, one block rides over the other, forming an overhanging FAULT SCARP. The angle of inclination of the fault plane from the horizontal is called the dip. The inclination of a fault plane is generally constant throughout the length of the fault, but there can be local variations in slope. In a STRIKE-SLIP FAULT the movement is horizontal, so no fault scarp is produced, although the FAULT LINE may be seen on the surface.

Fault scarp near Red Canyon Creek, Montana. (U.S. Geological Survey)

Fault scarp. FAULTS are produced through breaking or fracture of the surface ROCKS of the earth's CRUST as a result of stresses arising from tectonic movement. A NORMAL FAULT, one in which the earth movement is predominantly vertical, produces a steep fault scarp. A STRIKE-SLIP FAULT does not produce a fault scarp.

Fauna. Total animal POPULATION of a COUNTRY or REGION, from the largest creatures to the smallest. From the Latin word for "animals." See also FLORA.

Feldspar. Family name for a group of common MINERALS found in such ROCKS as GRANITE and composed of silicates of aluminum together with potassium, sodium, and calcium. Feldspars are the most abundant group of minerals within the earth's CRUST. There are many varieties of feldspar, distinguished by variations in chemistry and crystal structure. Although feldspars have some economic uses, their principal importance lies in their role as rock-forming minerals.

Fell. English word for an open grassy highland, such as a moor.

Felsic rocks. IGNEOUS ROCKS rich in potassium, sodium, aluminum, and SILICA, including GRANITES and related rocks.

Fen. Low-lying WETLAND; a BOG or MARSH.

Feng shui. Ancient Chinese philosophic system that ascribes good and bad qualities to the physical ENVIRONMENT and seeks to determine which LOCATIONS to choose, which to avoid, or how they might be modified to create a favorable set of conditions for human occupancy.

Fertility rate. DEMOGRAPHIC MEASURE of the average number of children per adult female in any given POPULATION. Religious beliefs, education, and other cultural considerations influence fertility rates. See also BIRTH RATE.

Fertilizer. Substance added to the SOIL to improve agricultural production. Plants need nitrogen for their growth, and this can be provided by organic fertilizers such as manure and compost. In high-income economies that practice commercial AGRICULTURE, farmers often use synthetic or manufactured inorganic fertilizers, which are produced in chemical plants and factories. Overapplication of inorganic fertilizer to cropland results in RUNOFF that produces excess nitrogen in STREAMS, LAKES, and, eventually, OCEANS.

Fetch. Distance along a large water surface over which a WIND of almost uniform direction and speed blows.

Feudalism. Social and economic system that prevailed in Europe before the INDUSTRIAL REVOLUTION. The land was owned and controlled by a minority comprising noblemen or lords; all other people were peasants or serfs, who worked as agricultural laborers on the lords' land. The peasants were not free to leave, or to do anything without their lord's permission. Other REGIONS such as China and Japan also had a feudal system in the past.

Fiord. See FJORD.

Firn. Intermediate stage between SNOW and glacial ice. Firn has a granular TEXTURE, due to compaction. Also called NÉVÉ.

Firth. Scottish word for a narrow ESTUARY. The Firth of Forth near Edinburgh is spanned by a famous steel railway BRIDGE constructed in 1890, which has been called Scotland's Eiffel Tower because of its engineering.

Fission, nuclear. Splitting of an atomic nucleus into two lighter nuclei, resulting in the release of neutrons and some of the binding ENERGY that held the nucleus together.

Fissure. Fracture or crack in ROCK along which there is a distinct separation.

Fjord. VALLEY produced at the COAST by a GLACIER that flowed to the SEA. The rising SEA LEVEL FLOODS the glacial TROUGH, producing a fjord.

Kenai Fjords National Park was established on Alaska's Kenai Peninsula— due south of Anchorage—in 1980 to protect the peninsula's many scenic fjords. (PhotoDisc)

The deep water enables ships to sail into fjords, and tourists enjoy the spectacular scenery. Fjord coasts include those of Norway, Alaska, the west coast of the South Island of New Zealand, and less-visited Chile and Antarctica. Also spelled "fiord."

Flash flood. Sudden rush of water down a STREAM CHANNEL, usually in the DESERT after a short but intense STORM. Other causes, such as a DAM failure, could lead to a flash flood.

Flood. Water overflowing a LEVEE and running out over the FLOODPLAIN when the volume of water in a STREAM becomes greater than the stream CHANNEL can contain.

Flood control. Attempts by humans to prevent flooding of STREAMS. Humans have consistently settled on FLOODPLAINS and DELTAS because of the fertile SOIL for AGRICULTURE, and attempts at flood control date back thousands of years. In strictly agricultural societies such as ancient Egypt, people built VILLAGES above the FLOOD levels, but transport and industry made riverside LOCATIONS desirable and engineers devised technological means to try to prevent flood damage. Artificial LEVEES, RESERVOIRS, and DAMS of ever-increasing size were built on RIVERS, as well as bypass CHANNELS leading to artificial floodplains. In many modern dam construction projects, the production of HYDRO-ELECTRIC POWER was more important than flood control. Despite modern TECHNOLOGY, floods cause the largest loss of human life of all natural disasters, especially in low-income countries such as Bangladesh.

Flood tide. Rising or incoming tide. Most parts of the world experience two flood TIDES in each 24-hour period. See also EBB TIDE.

Floodplain. Flat, low-lying land on either side of a STREAM, created by the DEPOSITION of ALLUVIUM from floods. Also called ALLUVIAL PLAIN.

Salt crust that has accumulated on a Southern California playa or lake bed as the result of evaporation. (U.S. Geological Survey)

Flora. All the plants of a COUNTRY or REGION, from the largest trees to the smallest mosses. From the Latin word for "flower." See also FAUNA.

Fluvial. Pertaining to running water; for example, fluvial processes are those in which running water is the dominant agent.

Mexican Hat, a gooseneck bend in the San Juan River in Utah's Goosenecks State Park is a spectacular example of fluvial erosion. (Corbis)

Fog. CLOUD in contact with the ground. Fog is generally a stratiform or layer cloud. Visibility is reduced to less than a half mile, making traveling hazardous.

Fog. (PhotoDisc)

Fog deserts. Coastal DESERTS where FOG is an important source of moisture for plants, animals, and humans. The fog forms because of a cold OCEAN CURRENT close to the SHORE. The Namib Desert of southwestern Africa, the west COAST of California, and the Atacama Desert of Peru are coastal deserts.

Föhn wind. WIND warmed and dried by descent, usually on the LEE side of a MOUNTAIN. In North America, these winds are called the CHINOOK.

Fold mountains. ROCKS in the earth's CRUST can be bent by compression, producing folds. The Swiss Alps are an example of complex FOLDING, accompanied by faulting. Simple upward folds are ANTICLINES, downward folds are SYNCLINES; but subsequent EROSION can produce LANDSCAPES with synclinal MOUNTAINS.

Folding. Bending of ROCKS in the earth's CRUST, caused by compression. The rocks are deformed, sometimes pushed up to form mountain RANGES. See also ISOCLINAL FOLDING.

Foliation. TEXTURE or structure in which MINERAL grains are arranged in parallel planes.

Food chain. Pattern found in nature by which organisms at one level provide food for those at the next level. The food chain represents the flow of ENERGY. At the bottom of the food chain are the producers—plants; all other levels are consumers. At the top of many food chains is the top CARNIVORE. An example of a simple food chain is: wolf eats rabbit, rabbit eats grass.

Food web. Complex network of FOOD CHAINS. Food chains are interconnected, because many organisms feed on a variety of others, and in turn may be eaten by any of a number of predators.

Forced migration. MIGRATION that occurs when people are moved against their will. The Atlantic slave trade is an example of forced migration. People were shipped from Africa to countries in Europe, Asia, and the New World as forced immigrants. Within the United States, some NATIVE AMERICANS were forced by the federal government to migrate to new reservations.

Ford. Short shallow section of a RIVER, where a person can cross easily, usually by walking or riding a horse. To cross a STREAM in such a manner.

Forest. Trees growing so closely together that their canopies meet or overlap. The existence of forest means abundant PRECIPITATION. In the TROPICS, tropical RAIN FOREST is found; in cooler midlatitude areas, DECIDUOUS FORESTS grow; in the cold northern REGIONS, BOREAL (CONIFEROUS) forests are evident. Where there is insufficient precipitation, WOODLAND or scrub is the dominant VEGETATION type. For centuries, forests have been removed for construction timbers, firewood, or farmland. Such DEFORESTATION can cause regional mudslides and FLOODS and, on a global scale, GLOBAL WARMING.

Formal region. Cultural REGION in which one trait, or group of traits, is uniform. LANGUAGE might be the basis of delineation of a formal cul-

Mormon temple in Salt Lake City, Utah, the center of world Mormonism and an example of a formal region. (Corbis)

tural region. For example, the Francophone region of Canada constitutes a formal region based on one single trait. One might also identify a formal Mormon region centered on the STATE of Utah, combining RELIGION and LANDSCAPE as defining traits. Cultural geographers generally identify formal regions using a combination of traits.

Fossil. Remains of ancient plants or animals preserved in layers of SEDIMENTARY ROCK. Most fossils belong to species that are now extinct.

Fossil beds. (PhotoDisc)

The study of fossils led to the development of the geologic time SCALE and the realization that the earth was billions of years old. Fossils now are dated by scientific methods such as CARBON DATING.

Fossil fuel. Deposit rich in hydrocarbons, formed from organic materials compressed in ROCK layers—COAL, OIL, and NATURAL GAS.

Fossil record. Fossil record provides evidence that addresses fundamental questions about the origin and history of life on the earth: When life evolved; how new groups of organisms originated; how major groups of organisms are related. This record is neither complete nor without biases, but as scientists' understanding of the limits and potential of the fossil record grows, the interpretations drawn from it are strengthened.

Fossilization. Processes by which the remains of an organism become preserved in the ROCK record.

Foucault's pendulum. Nineteenth century French physicist Jean-Bernard-Léon Foucault used a giant pendulum to demonstrate the ROTATION of the earth on its AXIS. While the pendulum swings to and fro in one plane, the earth rotates beneath it so the relative position changes. In the NORTHERN HEMISPHERE, a pendulum rotates CLOCKWISE because of the CORIOLIS EFFECT. Foucault also invented the gyroscope.

Fracture zones. Large, linear zones of the seafloor characterized by steep CLIFFS, irregular TOPOGRAPHY, and FAULTS; such zones commonly cross and displace oceanic RIDGES by faulting.

Free association. Relationship between sovereign NATIONS in which one nation—invariably the larger—has responsibility for the other nation's defense. The Cook Islands in the South Pacific have such a relationship with New Zealand.

Fresh water. Water with less than 0.2 percent dissolved SALTS, such as is found in most STREAMS, RIVERS, and LAKES.

Friction of distance. Distance is of prime importance in social, political, economic, and other relationships. Large distance has a negative effect on human activity. The time and cost of overcoming distance can be a deterrent to various activities. This has been called the friction of distance.

Frigid zone. Coldest of the three CLIMATE zones proposed by the ancient Greeks on the basis of their theories about the earth. There were two frigid zones, one around each POLE. The Greeks believed that human life was possible only in the TEMPERATE ZONE.

Fringing reef. Type of CORAL REEF formed at the SHORELINE, extending out from the land in shallow water. The top of the coral may be exposed at low TIDE.

Front. BOUNDARY between two AIR masses with different TEMPERATURE and moisture characteristics. When warm air moves in, a warm front is produced; when cold air moves in, a COLD FRONT is produced. Rain and changes in temperature and WIND direction accompany the passage of a front. A typical midlatitude CYCLONE, as it moves across North America from west to east, comprises a warm front followed by a cold front. Fronts can be stationary, when no movement is taking place, or occluded, when a cold front overtakes a warm front.

Frontier. Remote, sparsely populated REGION, which may hold potential for DEVELOPMENT, such as MINERAL deposits. Alaska might be regarded as the "last frontier" of the United States.

Frontier Thesis. Thesis first advanced by the American historian Frederick Jackson Turner, who declared that American history and the American character were shaped by the existence of empty, FRONTIER lands that led to exploration and westward expansion and DEVELOPMENT. The closing of the frontier occurred when transcontinental railroads linked the East and West Coasts and SETTLEMENTS spread across the United States. This thesis was used by later historians to explain the history of South Africa, Canada, and Australia. Critics of the Frontier Thesis point out that minorities and women were excluded from this view of history.

Frost. Thin white covering of ice crystals formed on the surface of objects and plants by the freezing of water vapor when the TEMPERATURE falls below 32 DEGREES Fahrenheit (0 degrees Celsius).

Frost wedging. Powerful form of PHYSICAL WEATHERING of ROCK, in which the expansion of water as it freezes in JOINTS or cracks shatters the rock into smaller pieces. Also known as frost shattering.

Fumarole. Crack in the earth's surface from which steam and other gases emerge. Fumaroles are found in volcanic areas and areas of GEOTHERMAL activity, such as Yellowstone National Park.

Functional region. Part of the earth's surface that is integrated or connected in a functional sense. A political unit such as a COUNTY, a METROPOLITAN statistical area, or an incorporated CITY is a functional region.

Funnel cloud. Narrow base of a TORNADO, between the bottom of a CUMULONIMBUS CLOUD and the ground, caused by the reduction of pressure at the center of the tornado. Devastation occurs as the funnel cloud moves rapidly along the ground. The cloud is dark because debris of all kinds has been sucked into the tornado. Most damage is caused by the strong swirling WINDS, but the low pressure at the center can cause buildings to explode. A WATERSPOUT also has a funnel cloud.

Fusion, nuclear. Collision and combining of two nuclei to form a single nucleus with less mass than the original nuclei, with a release of ENERGY equivalent to the mass reduction.

Fusion energy. Heat derived from the natural or human-induced union of atomic nuclei; in effect, the opposite of FISSION energy.

Gale. Strong WIND. On the BEAUFORT wind SCALE, gale force RANGES from 30 miles (50 km.) per hour (moderate) through fresh gale and strong gale, to a whole gale or STORM, when windspeeds are 48 to 55 knots (88 to 101 km.) per hour. At SEA, the progression is from blown sea spray to a foam-covered sea with very high WAVES. On land, a moderate gale means entire trees move; in a whole gale, trees are uprooted and considerable structural damage occurs.

Gall's projection. MAP PROJECTION constructed by projecting the earth onto a cylinder that intersects the sphere at 45 DEGREES north and 45 degrees south LATITUDE. The resulting map has less distortion of area than the more familiar CYLINDRICAL PROJECTION of Mercator. See also MAP.

Gangue. Apparently worthless ROCK or earth in which valuable gems or MINERALS are found.

Gap. Steep VALLEY or GORGE cut by a STREAM as it flows through an area of hard ROCK. In some cases, the stream stops flowing or disappears, usually because of stream capture. The LANDFORM feature is then called a WIND GAP.

Garigue. VEGETATION cover of small shrubs found in Mediterranean areas. Similar to the larger *maquis.*

Gas giant. Large planetary body that is primarily composed of hydrogen

Best known for its spectacular rings, Saturn is one of four gas giant planets in the Solar System. (PhotoDisc)

and helium, with minor amounts of other components; Jupiter, Saturn, Uranus, and Neptune are gas giants.

Gateway city. CITY whose physical LOCATION makes it a link between one COUNTRY and others, or between one REGION and others. A gateway city exercises control over a large area, because it commands the entry and exit rights and powers for a particular country or region. Most gateway cities are PORTS, many of which were formerly administrative centers for a colonial government. New York began as a small fur-trading outpost, but in the nineteenth century, it became a gateway for millions of immigrants from Europe to America. In colonial Brazil, Salvador was the gateway city through which more than three million slaves were brought from Africa to work on Portuguese-owned PLANTATIONS.

Gemstone. Any ROCK, MINERAL, or natural material that has the potential for use as personal adornment or ornament. Examples include diamonds, emeralds, rubies, and sapphires.

Gentrification. Phenomenon that occurs when the older housing stock of inner-CITY, working-class neighborhoods is purchased and renovated as a residential area for higher-income households. The new purchasers are attracted by the convenience of an inner-city LOCATION and lower prices, but gentrification displaces many of the older, original inhabitants.

Genus (plural, genera). Group of closely related species; for example, *Homo* is the genus of humans, and it includes the species *Homo sapiens* (modern humans) and *Homo erectus* (Peking Man, Java Man).

Geochronology. Study of the time SCALE of the earth; it attempts to develop methods that allow the scientist to reconstruct the past by dating events such as the formation of ROCKS.

Geodesy. Branch of applied mathematics that determines the exact positions of points on the earth's surface, the size and shape of the earth, and the variations of terrestrial GRAVITY and MAGNETISM.

Geoid. Figure of the earth considered as a MEAN SEA LEVEL surface extended continuously through the CONTINENTS.

Geologic map. MAP illustrating the age, structure, and DISTRIBUTION of ROCK units.

Geologic record. History of the earth and its life as recorded in successive layers of SEDIMENT and the FOSSIL specimens they contain.

Geologic terrane. Crustal block with a distinct group of ROCKS and structures resulting from a particular geologic history; assemblages of TERRANES form the CONTINENTS.

Geological column. Order of ROCK layers formed during the course of the earth's history.

Geomagnetic elements. MEASUREMENTS that describe the direction and intensity of the earth's MAGNETIC FIELD.

Geomagnetic poles. See MAGNETIC POLES.

Geomagnetism. External MAGNETIC FIELD generated by forces within the earth; this force attracts materials having similar properties, inducing them to line up (point) along field lines of force.

Geomorphic cycle. See CYCLE OF EROSION.

Geomorphology. Study of the origins of LANDFORMS and the processes of landform development.

Geophysics. Quantitative evaluation of ROCKS and surface features of the earth by electrical, gravitational, magnetic, radioactive, and elastic wave transmission and heat-flow techniques.

Geostationary orbit. ORBIT in which a SATELLITE appears to hover over one spot on the PLANET'S EQUATOR; this procedure requires that the orbit be high enough that its period matches the planet's rotational period, and have no inclination relative to the equator; for Earth, the ALTITUDE is 22,260 miles (35,903 km.).

Geostrophic. Force that causes directional change because of the earth's ROTATION.

Geotherm. Curve on a TEMPERATURE-depth graph that describes how temperature changes in the subsurface.

Geothermal. Pertaining to the heat of the interior of a PLANET.

Geothermal power. Power having its source in the earth's internal heat.

Geyser. Type of HOT SPRING that periodically erupts steam and hot water. Geysers are surface expressions of vast underground circulation sys-

One of the most famous geysers in the world is Yellowstone National Park's Old Faithful, which owes its nickname to the clocklike regularity with which it erupts. (Digital Stock)

tems, where constituents from underground ROCKS are dissolved in the hot fluids, carried to the surface, and deposited. The world's active thermal areas are natural laboratories where ORE-forming processes can be observed at first hand.

Glacial erratic. (U.S. Geological Survey)

Glacial erratic. ROCK that has been moved from its original position and transported by becoming incorporated in the ice of a GLACIER. Deposited in a new LOCATION, the rock is noteworthy because its geology is completely different from that of the surrounding rocks. Glacial erratics provide information about the direction of glacial movement and strength of the flow. They can be as small as PEBBLES, but the most interesting erratics are large boulders. Erratics become smoothed and rounded by the transport and EROSION.

Glaciation. This term is used in two senses: first, in reference to the cyclic widespread growth and advance of ICE SHEETS over the polar and high- to mid-LATITUDE REGIONS of the CONTINENTS; second, in reference to the effect of a GLACIER on the TERRAIN it transverses as it advances and recedes.

Mount Shuksan, in northern Washington's North Cascades National Park, has nine major glaciers, which have sculpted it to resemble peaks in the Swiss Alps. (Corbis)

Glacier. Tightly packed snowmass that grows larger as it receives more PRECIPITATION and moves forward—often with enough power to re-shape land formations.

Glaciology. Scientific study of GLACIERS and ice.

Global Positioning System (GPS). Group of SATELLITES that ORBIT Earth every twenty-four hours, sending out signals that can be used to locate PLACES on Earth and in near-Earth orbits.

Global warming. Trend of Earth CLIMATES to grow increasingly warm as a result of the GREENHOUSE EFFECT. One of the most dramatic effects of global warming is the melting of the POLAR ICE CAPS and a consequent rise in the level of the world's OCEANS.

Gondwanaland. Hypothesized ancient CONTINENT in the SOUTHERN HEMISPHERE that geologists theorize broke into at least two large segments; one segment became India and pushed northward to collide with the Eurasian LANDMASS, while the other, Africa, moved westward. Australia and Antarctica were also part of Gondwanaland.

Gorge. Steeply walled CANYON or section of a canyon.

The Columbia River Gorge is a break in the Cascade Range through which the Columbia River passes. The Lewis and Clark Expedition reached the gorge in 1806. (PhotoDisc)

Graben. Roughly symmetrical crustal depression formed by the lowering of a crustal block between two NORMAL FAULTS that slope toward each other.

Granite. Coarse-grained, commonly light-colored PLUTONIC IGNEOUS ROCK composed primarily of two FELDSPARS (plagioclase and ortho-clase) and QUARTZ, with variable amounts of dark MINERALS.

Stone Mountain in northwestern Georgia, has a monumental relief carved into its granite northern face. (PhotoDisc)

Granules. Small grains or pellets.

Grassland. Two of Earth's major BIOMES, grasslands cover about a quarter of the world's land surface. Because the temperate grassland regions constitute the PLANET's richest SOILS, they are intensely farmed and grazed, and only small patches of natural grassland remain. Tropical grasslands are usually known as SAVANNA.

Gravimeter. Device that measures the attraction of GRAVITY.

Gravitational differentiation. Separation of MINERALS, elements, or both as a result of the influence of a gravitational field wherein heavy phases sink or light phases rise through a melt.

Gravity. Natural attractive force exerted by Earth on objects on or near its surface.

Great circle. Largest circle that goes around a sphere. On the earth, all lines of LONGITUDE are parts of great circles; however, the EQUATOR is the only line of LATITUDE that is a great circle.

Green mud. SOILS that develop under conditions of excess water, or waterlogged soils, can display colors of gray to blue to green, largely because of chemical reactions involving iron. Fine CLAY soils and muds in areas such as BOGS or ESTUARIES can be called green mud. This soil-forming process is called gleization.

Greenhouse effect. Trapping of the SUN's rays within the earth's ATMO-SPHERE, with a consequence rise in TEMPERATURES that leads to GLOBAL WARMING.

Greenhouse effect gets its name because clouds and gases of the lower atmosphere trap surface radiation in a manner similar to that of buildings called greenhouses, such as this greenhouse for tree seedlings. (PhotoDisc)

Greenhouse gas. Atmospheric gas capable of absorbing electromagnetic radiation in the infrared part of the spectrum.

Greenwich mean time. Also known as universal time, the solar mean time on the MERIDIAN running through Greenwich, England—which is used as the basis for calculating time throughout most of the world.

Grid. Pattern of horizontal and vertical lines forming squares of uniform size.

Gross domestic product. Value representing the total value of all goods, food, MINERALS, and services produced in a particular COUNTRY in one year. This total value usually is divided by the total POPULATION of the country, so that the figure given is the gross domestic product (GDP) per capita. The GDP is often used to compare the standard of living in different countries. For high-income economies, some economists and other researchers prefer to use the gross national product per capita.

Gross migration. Total number of migrants moving into and out of a RE-GION. The balance between these two MIGRATION streams is called NET MIGRATION.

Groundwater. Water that occurs beneath the surface of the earth, as opposed to SURFACE WATER that occurs in RIVERS and LAKES. Most groundwater comes from PRECIPITATION, when water percolates through soil or ROCK until it is stopped by an impervious rock layer called an aquiclude. The rocks that store groundwater in this way are called AQUIFERS. WELLS are drilled to pump groundwater to the surface for IRRIGATION and for human consumption. Groundwater accounts for about 0.6 percent of the earth's total HYDROSPHERE.

Groundwater movement. Flow of water through the subsurface, known as groundwater movement, obeys set principles that allow hydrologists to predict flow directions and rates.

Groundwater recharge. Water that infiltrates from the surface of the earth downward through SOIL and ROCK pores to the WATER TABLE, causing its level to rise.

Growth pole. LOCATION where high-growth economic activity is deliberately encouraged and promoted. Governments often establish growth poles by creating industrial parks, open cities, special economic zones, new TOWNS, and other incentives. The plan is that the new industries will further stimulate economic growth in a cumulative trend. Automobile plants are a traditional form of growth industry but have been overtaken by high-tech industries and BIOTECHNOLOGY. In France, the term "technopole" is used for a high-tech growth pole. A related concept is SPREAD EFFECTS.

Guano. Fossilized bird excrement, found in abundance on some COASTS or ISLANDS, notably Nauru in the Pacific.

Guest workers. People who migrate temporarily to another COUNTRY for jobs. Much of the money they earn is sent back to families in the HOMELAND. Guest workers are a form of economic migrants, but the emphasis is on the temporary nature of their residence in the new country. After World War II, a shortage of industrial and factory workers led Germany to invite guest workers from Greece, Italy, Yugoslavia, and Turkey to provide labor in the newly rebuilt factories, or to fill low-paid positions. France has many guest workers from northern African countries. Guest workers pose social problems in the new country. Their presence is sometimes resented by nationals. Most guest workers are young men, which can lead to social problems with prostitution, for example. Guest workers tend to form residential ENCLAVES in low-rent areas of a CITY, creating a kind of ghetto.

Gulf. Large OCEAN INLET. "Gulf" is not a precise term but it is usually applied to inlets larger than BAYS.

Guyot. Drowned volcanic ISLAND with a flat top caused by WAVE EROSION or coral growth. A type of SEAMOUNT.

Gyre. Large semiclosed circulation patterns of OCEAN CURRENTS in each of the major OCEAN BASINS that move in opposite directions in the Northern and Southern hemispheres. There are five gyres in the world's oceans.

Haff. Term used for various WETLANDS or LAGOONS located around the southern end of the Baltic Sea, from Latvia to Germany. Offshore BARS of SAND and shingle separate the haffs from the open SEA. One of the largest is the Stettiner Haff, which covers the BORDER REGION between Germany and Poland and is separated from the Baltic by the low-lying ISLAND of Usedom. The Kurisches Haff (in English, the Courtland Lagoon) is located on the Lithuanian border.

Hamlet. Loose term for a human SETTLEMENT that would be considered smaller than a VILLAGE.

Harbor. INLET, or protected body of water, that serves as an anchorage for shipping or small boats.

Harmonic tremor. Type of EARTHQUAKE activity in which the ground undergoes continuous shaking in response to subsurface movement of MAGMA.

Headland. Elevated land projecting into a body of water.

Headwaters. Source of a RIVER. Also called headstream.

Heat index. Measure combining TEMPERATURE and RELATIVE HUMIDITY to indicate an apparent or sensible temperature, which is a guide to the danger of overexertion in certain WEATHER conditions.

Heat sink. Term applied to Antarctica, whose cold CLIMATE causes warm AIR masses flowing over it to chill quickly and lose ALTITUDE, affecting the entire world's WEATHER.

Hemisphere. Geometrical term for half of a sphere. All spherical celestial objects, such as PLANETS and stars, have NORTHERN and SOUTHERN HEMISPHERES divided by the bodies' EQUATORS. Hemisphere defined by MERIDIANS are more arbitrary. The earth is generally regarded as being divided into EASTERN and WESTERN HEMISPHERES, but their REGIONS are not precisely defined.

Heterosphere. Major realm of the ATMOSPHERE in which the gases hydrogen and helium become predominant.

High-frequency seismic waves. EARTHQUAKE WAVES that shake the ROCK through which they travel most rapidly.

High island. See CONTINENTAL ISLAND.

Hill. Term loosely applied to an elevated mass of land that would be considered smaller than a MOUNTAIN. In contrast to a PEAK, a hill usually has a smooth SUMMIT.

Hillock. Small natural HILL. A similar but smaller feature is a hummock. There is no standard definition for these terms.

Hinterland. Area that surrounds a CITY and relies on the city for goods and services. The city, in turn, may draw RESOURCES from its hinterland. From the German word for "country behind."

Histogram. Bar graph in which vertical bars represent frequency and the horizontal axis represents categories. A POPULATION PYRAMID, or age-sex pyramid, is a histogram, as is a CLIMOGRAPH.

Historical inertia. Term used by economic geographers when heavy industries, such as steelmaking and large manufacture, that require huge capital investments in land and plant continue in operation for long periods, even after they become out of date, uncompetitive, or obsolete.

Hoar frost. Similar to DEW, except that moisture is deposited as ice crystals, not liquid dew, on surfaces such as grass or plant leaves. When moist AIR cools to saturation level at TEMPERATURES below the freezing point, CONDENSATION occurs directly as ice. Technically, hoar frost is not the same as frozen dew, but it is difficult to distinguish between the two.

Hogback. Steeply sloping homoclinal RIDGE, with a slope of 45 DEGREES or more. The angle of the slope is the same as the dip of the ROCK STRATA. These LANDFORMS develop in REGIONS where the underlying rocks, usually SEDIMENTARY, have been folded into anticlinal ridges and synclinal VALLEYS. Differential EROSION causes softer rock layers to wear away more rapidly than the harder layers of rock that form the hogback ridge. A similar feature with a gentler slope is called a CUESTA.

Holocene. Name for the current, or modern, geological EPOCH. It began around ten thousand years ago, after the PLEISTOCENE.

Homeland. CULTURE REGION to which a group of humans have an emotional attachment. ETHNIC GROUPS are usually identified with a homeland, which comprises a physical LANDSCAPE and the historical events that occurred there.

Homosphere. Lower part of the earth's ATMOSPHERE. In this area, 60 miles (100 km.) thick, the component gases are uniformly mixed together, largely through WINDS and turbulent AIR CURRENTS. Above the homosphere is the REGION of the atmosphere called the HETEROSPHERE. There, the individual gases separate out into layers on the basis of their molecular weight. The lighter gases, hydrogen and helium, are at the top of the heterosphere.

Hook. A long, narrow deposit of SAND and SILT that grows outward into the OCEAN from the land is called a SPIT or sandspit. A hook forms

Cape Cod, Massachusetts, is the most famous spit and hook in the United States. (PhotoDisc)

when currents or WAVES cause the deposited material to curve back toward the land. Cape Cod is the most famous spit and hook in the United States.

Horizon, true. GREAT CIRCLE of the celestial sphere. It is formed by the intersection of the celestial sphere and a plane through the center of the earth, and is perpendicular to the zenith-nadir line. The true horizon is not the same as the visible HORIZON, which is the line where earth and sky appear to a viewer to meet. Also known as rational horizon.

Horizon, visible. Line where the sky seems to meet the SEA or land. The ALTITUDE of the observer affects the distance between that person and the visible horizon: A person standing on a MOUNTAIN perceives the horizon as being a much greater distance away than a person at SEA LEVEL. Also called sensible or rational horizon.

Horse latitudes. Parts of the OCEANS from about 30 to 35 DEGREES north or south of the EQUATOR. In these latitudes, AIR movement is usually light WINDS, or even complete calm, because there are semipermanent high-pressure cells called ANTICYCLONES, which are marked by dry subsiding air and fine clear WEATHER. The atmospheric circulation of an anticyclone is divergent and CLOCKWISE in the NORTHERN HEMISPHERE, so to the north of the horse latitudes are the westerly winds and to the south are the northeast TRADE WINDS. In the SOUTHERN HEMISPHERE, the circulation is reversed, producing the easterly winds and the southeast trade winds. It is believed that the name originated because when ships bringing immigrants to the Americas were becalmed for any length of time, horses were thrown overboard because they required too much FRESH WATER. Also called the CALMS OF CANCER.

Horst. FAULT block or piece of land that stands above the surrounding land. A horst usually has been uplifted by tectonic forces, but also could have originated by downward movement or lowering of the adjacent lands. Movement occurs along the parallel faults on either side of a horst. If the land is downthrown instead of uplifted, a VALLEY known as a GRABEN is formed. "Horst" comes from the German word for horse, because the flat-topped feature resembles a vaulting horse used in gymnastics.

Horticulture. Cultivation of plants in gardens or orchards to produce food for one's own consumption or for sale. Horticulture is a form of commercial AGRICULTURE and is usually found near large cities, where there is a ready market for fresh produce. "Market gardening" is a similar term.

Hot spot. PLACE on the earth's surface where heat and MAGMA rise from deep in the interior, perhaps from the lower MANTLE. Erupting VOLCANOES may be present, as in the formation of the Hawaiian Islands. More commonly, the heat from the rising magma causes GROUNDWATER to form HOT SPRINGS, GEYSERS, and other thermal and HYDRO-

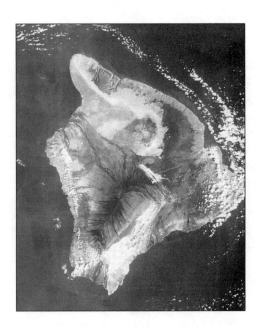

Landsat image of the big island of Hawaii, which rests on a plate that is moving across a hot spot. (U.S. Geological Survey)

thermal features. Yellowstone National Park is located on a hot spot. Also known as a MANTLE PLUME.

Hot spring. SPRING where hot water emerges at the earth's surface. The usual cause is that the GROUNDWATER is heated by MAGMA. A GEYSER is a special type of hot spring at which the water heats under pressure

Mammoth Hot Springs in Yellowstone National Park comprises about seventy separate springs, which maintain water temperatures between 60 and 175 degrees Fahrenheit (15 to 80 degrees Celsius). (Digital Stock)

and that periodically spouts hot water and steam. Old Faithful is the best known of many geysers in Yellowstone National Park. In some countries, GEOTHERMAL ENERGY from hot springs is used to generate electricity. Also called thermal spring.

Huerta. Irrigated orchard or agricultural parcel of land in southern Spain. The MEDITERRANEAN CLIMATE, IRRIGATION, and intensive labor make the Spanish huertas productive. Typical crops include grains such as corn and wheat, citrus, peaches, nuts, grapes, and dates; beef cattle are raised also. If the irrigated land is used mainly for the production of fodder crops, it may be called a vega.

Humid-midlatitude. Land area with average TEMPERATURE of the coldest month less than 64 DEGREES Fahrenheit (18 degrees Celsius) but at least eight months with average monthly temperatures greater than 50 degrees Fahrenheit (10 degrees Celsius); this area has no dry season.

Humidity. Water vapor in the earth's ATMOSPHERE. Concentrated in the lower 1 mile (1.6 km.) of the TROPOSPHERE. It may be measured as ABSOLUTE HUMIDITY (in grams per cubic meter), as specific humidity (in grams per kilogram of AIR), or, most commonly, as RELATIVE HUMIDITY—a percentage that represents the amount of water vapor in the air at a given TEMPERATURE, compared with the amount the air could contain if it were saturated. High humidity causes discomfort because evaporative cooling is hampered.

Hummock. See HILLOCK.

Hummocky. TOPOGRAPHY characterized by a slope composed of many irregular mounds (hummocks) that are produced during sliding or flowage movements of earth and ROCK.

Humus. Uppermost layer of a SOIL, containing decaying and decomposing organic matter such as leaves. This produces nutrients, leading to a fertile soil. Tropical soils are low in humus, because the rate of decay is so rapid. Soils of GRASSLANDS and DECIDUOUS FOREST develop thick layers of humus. In a SOIL PROFILE, the layer containing humus is the O Horizon.

Hunting and gathering. Preagricultural ECONOMY based on finding and harvesting edible forms of wildlife and plants.

Hurricane. North American term for a tropical rotating STORM with low pressure in the center and WIND speeds in excess of 74 miles (64 knots/119 km.) per hour. Elsewhere called a TROPICAL CYCLONE or TYPHOON. Hurricanes develop near the EQUATOR over tropical OCEANS, usually in the summer when the water is warmest. In general, the path in the NORTHERN HEMISPHERE is to the northwest, and in the SOUTHERN HEMISPHERE to the southwest. The diameter of a hurricane can vary from 50 to 500 miles (80-800 km.), with wind speed increasing toward the center. At the center is the quiet EYE, a zone 10 to 25 miles (16-40 km.) in diameter, where there is no wind, pressure is extremely low, and the sky is clear. Surrounding the eye is the eye wall, where tall

Hurricane winds can reach strengths that not only bend trees but pull them out of the ground by their roots. (PhotoDisc)

CUMULONIMBUS CLOUDS swirl upward, rain falls heavily, and wind speeds are greatest. A hurricane has a life of about one week, although it loses ENERGY as soon as it crosses from ocean to land. Wind damage to property is considerable in a hurricane, but the greatest loss of life is caused by the resulting flooding. Torrential rain leads to swollen RIVERS; another factor is the STORM SURGE that originates when winds raise the ocean level to as much as 22 feet (7 meters) above the normal high-tide level. Parts of low-lying Bangladesh and India have suffered huge losses of life from these storms, which are called CYCLONES there. Hurricanes begin as TROPICAL DEPRESSIONS, which have a wind speed of up to 37 miles (61 km.) per hour. When wind speeds reach 38 to 70 miles (63 to 117 km.) per hour, the storm is classified as a TROPICAL STORM and a name is assigned. Tropical storms are watched carefully because they can develop into full hurricanes. It is thought that GLOBAL WARMING will lead to more frequent hurricanes, occurring in a wider area.

Hydroelectric power. Electricity generated when falling water turns the blades of a turbine that converts the water's potential ENERGY to mechanical energy. Natural WATERFALLS can be used, but most hydroelectric power is generated by water from DAMS, because the flow of water from a dam can be controlled. Hydroelectric generation is a RENEWABLE, clean, cheap way to produce power, but dam construction inundates land, often displacing people, who lose their homes, VILLAGES, and farmland. Aquatic life is altered and disrupted also; for example, Pacific salmon cannot return upstream on the Columbia River to their spawning REGION. In a few coastal PLACES, TIDAL ENERGY is used to generate hydroelectricity; La Rance in France is the oldest successful tidal power plant.

Hydrography. Surveying of underwater features or those parts of the earth that are covered by water, especially OCEAN depths and OCEAN CURRENTS. Hydrographers make MAPS and CHARTS of the ocean floor and COASTLINES, which are used by mariners for NAVIGATION. For centuries, mariners used a leadline, a long rope with a lead weight at the bottom. The line was thrown overboard and the depth of water measured. The unit of MEASUREMENT was FATHOMS (6 feet/1.8 meters), which is one-thousandth of a NAUTICAL MILE. The invention of sonar (underwater echo sounding) has enabled mapping of large areas, and hydrographers currently use both television cameras and SATELLITE data.

Hydrologic cycle. Continuous circulation of the earth's HYDROSPHERE, or waters, through EVAPORATION, CONDENSATION, and PRECIPITATION. Other parts of the hydrologic cycle include RUNOFF, INFILTRATION, and TRANSPIRATION.

Hydrology. Scientific study of all aspects of water, especially the operation of the various parts of the HYDROLOGIC CYCLE. Hydrologists are concerned with water at or near the earth's surface; oceanographers study the waters of the OCEAN. To study the relationship between water and the living ENVIRONMENT, a hydrologist needs to know botany, geology, chemistry, SOIL science, and computer modeling. Hydrologists carry out research related to DAM construction; FLOOD CONTROL; agricultural developments, including irrigated farming; HYDROELECTRIC POWER generation; ACID RAIN and its impacts; disposal of solid and liquid wastes; and recreational facilities. SATELLITE imagery is used widely in modern hydrology.

Hydrosphere. All the waters of the earth, which comprise more than 300 million cubic miles (approximately 1.3 billion cubic km.). More than 97 percent of the hydrosphere is contained in the OCEANS; ICE SHEETS and GLACIERS make up more than 2 percent of the total. Freshwater LAKES and RIVERS account for only 0.0091 percent of the earth's hydrosphere.

Hydrostatic pressure. Pressure imposed by the weight of an overlying column of water.

Hydrothermal. Characterizing any process involving hot GROUNDWATER or MINERALS formed by such processes.

Hydrothermal vents. Areas on the OCEAN floor, typically along FAULT LINES or in the vicinity of undersea VOLCANOES, where water that has percolated into the ROCK reemerges much hotter than the surround-

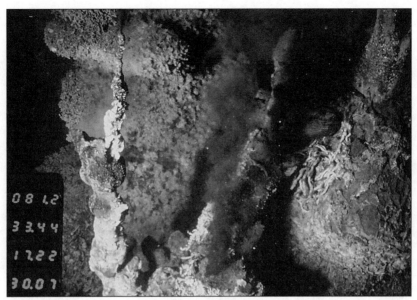

Hydrothermal vent with "black smokers," where plumes of superheated water meet colder water and dark minerals begin to precipitate out and form "chimneys." (National Oceanic and Atmospheric Administration)

ing water; such heated water carries various dissolved MINERALS, including metals and sulfides.

Hyetograph. Chart showing the DISTRIBUTION of RAINFALL over time. Typically, a hyetograph is constructed for a single STORM, showing the amount of total PRECIPITATION accumulating throughout the period. A hyetograph shows how rainfall intensity varies throughout the duration of a storm.

Hygrogram. Record made by a HYGROGRAPH. Under natural conditions, a hygrogram shows the increase of HUMIDITY in the early morning and the decrease each DAY as the TEMPERATURE rises. If a constant humidity needs to be maintained, a hygrograph is a way to monitor this factor.

Hygrograph. HYGROMETER that produces a record of RELATIVE HUMIDITY in the form of a graph or chart. A pen moves over a graph paper that is attached to a rotating cylinder. The cylinder can rotate once in 24 hours or once a week. The recording of HUMIDITY is based on the property of human hair to increase in length as humidity rises (known in daily life as a "bad hair day"). As the hair in the hygrograph absorbs moisture or dries out, it causes the pen to change position on the graph paper. Museums, especially art museums, often have a hygrograph to check on conditions for the delicate objects they display. When there are two pens, the second recording TEMPERATURE, the instrument is called a thermohygrograph.

Hygrometer. Instrument for measuring the RELATIVE HUMIDITY of AIR, or the amount of water vapor in the ATMOSPHERE at any time.

Hygrophyte. Plant that is adapted to living in wet conditions, usually with its roots permanently wet. Hygrophytic trees, such as the SWAMP cypress, have trunks that flare out into buttresses at the base to stabilize the tall tree. Other plants, like reeds and water lilies, have soft stems that can sway with the water movement.

Hypocenter. Central underground LOCATION of an earth tremor; also called the focus.

Hypsometer. Instrument used for measuring ALTITUDE (height above SEA LEVEL), using boiling water that circulates around a THERMOMETER. Since ATMOSPHERIC PRESSURE falls with increased altitude, the boiling point of water is lower. The hypsometer relies on this difference in boiling point to calculate ELEVATION. A more common instrument for measuring altitude is the ALTIMETER.

Ice age. PERIOD of geologic time when large parts of the earth's land surface were covered with ice and GLACIERS, because of a lowering of atmospheric TEMPERATURE. There have been several ice ages throughout Earth's history. The most recent began around two million years ago. See PLEISTOCENE.

Ice blink. Bright, usually yellowish-white glare or reflection on the underside of a CLOUD layer, produced by light reflected from an ice-covered surface such as pack ice. A similar phenomenon of reflection from a snow-covered surface is called snow blink.

Ice-cap climate. Earth's most severe CLIMATE, where the mean monthly TEMPERATURE is never above 32 DEGREES Fahrenheit (0 degrees Celsius). This climate is found in Greenland and Antarctica, which are high PLATEAUS, where KATABATIC WINDS blow strongly and frequently. At these high LATITUDES, INSOLATION (SOLAR ENERGY) is received for a short period in the summer months, but the high reflectivity of the ice and SNOW means that much is reflected back instead of being absorbed by the surface. No VEGETATION can grow, because the LANDSCAPE is permanently covered in ice and snow. Because AIR temperatures are so cold, PRECIPITATION is usually less than 5 inches (13 centimeters) annually. The POLES are REGIONS of stable, high-pressure air, where dry conditions prevail, but strong winds that blow the snow around are common. In the KOEPPEN CLIMATE CLASSIFICATION, the ice-cap climate is signified by the letters *EF.*

Ice caps. Small ICE SHEETS circular in shape covering areas of less than 19,300 square miles (50,000 sq. km.). See also POLAR ICE CAP.

Ice field. Similar to an ICE CAP, but elongated instead of forming a dome shape. Isolated PEAKS or RIDGES can protrude above the ice field. The best and largest example is the ice field in Patagonia in the Andes mountains of Argentina and Chile.

Ice sheet. Huge CONTINENTAL GLACIER. The only ice sheets remaining cover most of Antarctica and Greenland. At the peak of the last ICE AGE, around eighteen thousand years ago, ice covered as much as one-third of the earth's land surfaces. In the NORTHERN HEMISPHERE, there were two great ice sheets—the Laurentide ice sheet, covering North America, and the Scandinavian ice sheet, covering northwestern Europe and Scandinavia.

Ice shelf. Portion of an ICE SHEET extending into the OCEAN.

Ice storm. STORM characterized by a fall of freezing rain, with the formation of glaze on Earth objects.

Iceberg. Large mass of freshwater ice floating in the OCEAN, having broken off (calved) from the SNOUT of a GLACIER or the edge of an ICE SHEET. CALVING produces tens of thousands of icebergs each year around the margins of Greenland and Antarctica during the warmest summer months. Icebergs vary in height from a few feet to the height of a ten-story building and can persist for years. Depending on the shape of the iceberg, 80 to 90 percent of its total mass is submerged. Icebergs are moved by WAVES, WINDS, and OCEAN CURRENTS. They can be eroded by waves; more commonly, they melt as they move into warmer waters. Icebergs from Greenland were observed as far south as Bermuda early in the twentieth century. In the North Atlantic Ocean, icebergs from western Greenland are moved south by the Labrador Current and enter shipping lanes, where they pose a severe danger to vessels on the busy route between North America and Europe. When the steamship *Titanic*, a supposedly unsinkable vessel, collided with an iceberg in 1912, it sank so quickly that fifteen hundred passengers and crew members were drowned or perished in the icy waters. Today, radar and sonar can give early warning of iceberg danger to ships.

Giant iceberg. (PhotoDisc)

Oceanographers, the U.S. Coast Guard, and mariners monitor and track icebergs that approach shipping lanes.

Icefoot. Long, tapering extension of a GLACIER floating above the seawater where it enters the OCEAN. Eventually, it breaks away and forms an ICEBERG.

Igneous. From the Latin *ignis* (fire), a term referring to ROCKS formed from the molten state or to processes that form such rocks.

Igneous rock. ROCKS formed when molten material or MAGMA cools and crystallizes into solid rock. The type of rock varies with the composition of the magma and, more important, with the rate of cooling. Rocks that cool slowly, far beneath the earth's surface, are igneous IN-TRUSIVE ROCKS. These have large crystals and coarse grains. GRANITE is the most typical igneous intrusive rock. When cooling is more rapid, usually closer to or at the surface, finer-grained igneous EXTRUSIVE ROCKS such as rhyolite are formed. If the magma flows out to the surface as LAVA, it may cool quickly, forming a glassy rock called obsidian. If there is gas in the lava, rocks full of holes from bubbles of escaping gases form; PUMICE and BASALT are common igneous extrusive rocks.

Immigration. Moving of new residents into an area on a permanent basis. The United States was the destination of many twentieth century immigrants. See also EMIGRATION.

Impact crater. Generally circular depression formed on the surface of a

The Barringer Meteor Crater in northern Arizona was the first meteor-impact site identified on Earth. Estimated to be more than twenty-five thousand years old, the crater is about six hundred feet (180 meters) deep and about 3,800 feet (1.2 km.) in diameter. (PhotoDisc)

PLANET by the impact of a high-velocity projectile such as a METEORITE, ASTEROID, or COMET.

Impact volcanism. Process in which major impact events produce huge CRATERS along with MAGMA RESERVOIRS that subsequently produce volcanic activity. Such cratering is clearly visible on the MOON, Mars, Mercury, and probably Venus. It is assumed that Earth had similar craters, but EROSION has erased most of the evidence.

Imperialism. Acquisition and retention of a colonial empire. Ancient empires included the Greek and Roman empires. More recently, empires in Europe, the Americas, and Africa were controlled by such European powers as Spain, Great Britain, France, and Russia.

Impervious rock. Also known as impermeable rock, materials through which water cannot pass. ROCKS through which water can pass are called pervious. Solid or massive GRANITE, for example, is impervious. Nevertheless, a granite outcrop may be pervious because of the presence of small cracks called JOINTS, or because of FISSURES in the rock. Water could pass through the outcrop along these openings. Most CLAYS are impervious.

Import substitution. Economic process in which domestic producers manufacture or supply goods or services that were previously imported or purchased from overseas and foreign producers.

Index fossil. Remains of an ancient organism that are useful in establishing the age of ROCKS; index fossils are abundant and have a wide geographic DISTRIBUTION, a narrow stratigraphic RANGE, and a distinctive form.

Indian summer. Short period, usually not more than a week, of unusually warm WEATHER in late October or early November in the NORTHERN HEMISPHERE. Before the Indian summer, TEMPERATURES are cooler and there can be occurrences of FROST. Indian summer DAYS are marked by clear to hazy skies and calm to light WINDS, but nights are cool. The weather pattern is a high-pressure cell or ridge located for a few days over the East Coast of North America. The name originated in New England, referring to the practice of NATIVE AMERICANS gathering foods for winter storage over this brief spell. Similar weather in England is called an Old Wives' summer.

Indigenous people. Native inhabitants of a REGION; the aboriginal peoples.

Industrial Revolution. Change of a society from a RURAL and agricultural lifestyle to one in which most people earn their living in the industrial or secondary sector of the ECONOMY. MIGRATION from rural VILLAGES to URBAN SETTLEMENTS accompanies this change. The first Industrial Revolution began in England in the early eighteenth century. Technological advances in iron smelting, and later steel production, were accompanied by the invention of the steam engine. This provided a source of power for many new types of machinery in spinning and weaving and the locomotive and related industries. The Industrial Revolution spread from Great Britain to the CONTINENT of Europe and, in the late nineteenth century, to the United States.

Industrialization. Change from an agricultural society or agricultural ECONOMY to one that derives most of its income from industrial production. The INDUSTRIAL REVOLUTION began in Great Britain in the eighteenth century and spread to many other countries. URBAN SETTLEMENTS grew dramatically in size as a result of the demand for labor in industrial establishments.

Infant mortality. DEMOGRAPHIC MEASURE calculated as the number of deaths in a year of infants, or children under one year of age, compared with the total number of live births in a COUNTRY for the same year. Low-income countries have high infant mortality rates, more than one hundred infant deaths per thousand.

Infauna. Organisms that live in the seafloor.

Infiltration. Movement of water into and through the SOIL.

Informal economy. Form of employment whereby a person sells goods or services without a government license, often on the streets. This is especially common in low-income economies and in URBAN AREAS where unemployment is high. Recent immigrants often resort to this means of livelihood.

Informal sector. Economic activities conducted without official regulation or control. Street vendors who operate without a permit are part of the informal sector, as are street performers and beggars. See also INFORMAL ECONOMY.

Informal settlements. See SQUATTER SETTLEMENTS.

Infrastructure. Man-made bases of a society, such as road networks, power lines, airports, schools, hospitals, railroads, and police services.

Initial advantage. In terms of economic DEVELOPMENT, not all LOCATIONS are suited for profitable investment. Some locations offer initial advantages, including an existing skilled labor pool, existing consumer markets, existing plants, and situational advantages. These advantages can also lead to clustering of a number of industries at a particular location and to further economic growth, which will provide the preconditions of initial advantage for further economic development.

Inlet. Any recess along a SHORELINE of a larger body of water. Specific terminology is not precise, but a BAY is generally larger than a COVE, and a GULF is larger than both.

Inlier. REGION of old ROCKS that is completely surrounded by younger rocks. These are often PLACES where ORES or MINERALS are found in commercial quantities.

Inner core. The innermost layer of the earth; the inner core is a solid ball with a radius of about 900 miles.

Inselberg. Exposed rocky HILL in a DESERT area, made of resistant ROCKS, rising steeply from the flat surrounding countryside. There are many

Uluru, or Ayers Rock, in Australia is perhaps the world's best-known example of an inselberg. (Digital Stock)

inselbergs in Africa, but Uluru (Ayers Rock) in Australia is possibly the most famous inselberg. The word is German for "island mountain."

Insolation. ENERGY received by the earth from the SUN, which heats the earth's surface. The average insolation received at the top of the earth's ATMOSPHERE at an average distance from the Sun is called the SOLAR CONSTANT. Insolation is predominantly shortwave radiation, with wavelengths in the RANGE of 0.39 to 0.76 micrometers, which corresponds to the visible spectrum. Less than half of the incoming SOLAR ENERGY reaches the earth's surface—insolation is reflected back into space by CLOUDS; smaller amounts are reflected back by surfaces, absorbed, or scattered by the atmosphere. Insolation is not distributed evenly over the earth, because of Earth's curved surface. Where the rays are perpendicular, at the SUBSOLAR POINT, insolation is at the maximum. The word is a shortened form of incoming (or intercepted) SOLAR RADIATION.

Insular climate. Island climates are influenced by the fact that no PLACE is far from the SEA. Therefore, both the DIURNAL (daily) TEMPERATURE RANGE and the annual temperature range are small.

Insurgent state. STATE that arises when an uprising or guerrilla movement gains control of part of the territory of a COUNTRY, then establishes its own form of control or government. In effect, the insurgents create a state within a state. In Colombia, for example, the government and armed forces have been unable to control several REGIONS where insurgents have created their own domains. This is generally related to coca growing and the production of cocaine. Civilian farmers are unable to resist the drug-financed "armies."

Intensive subsistence agriculture. Practice whereby a small area of agricultural land produces an abundant crop, usually as a result of intensive human labor and the application of FERTILIZER. Countries of Asia where wet rice is grown practice intensive subsistence agriculture. The POPULATION pressure on the land is high, but the combination of high

TEMPERATURE, abundant RAINFALL, rich SOILS, and the productivity of rice as a crop enable large numbers of people to exist in this way. Terracing of hillsides to increase the available farming land is typical in these areas.

Intercropping. Growing of more than one crop in the same agricultural plot or field. Intercropping is commonly practiced by shifting cultivators. SUSTAINABLE AGRICULTURE uses intercropping as an alternative to pesticides.

Interfluve. Higher area between two STREAMS; the surface over which water flows into the stream. These surfaces are subject to RUNOFF and EROSION by RILL action and GULLYING. Over time, interfluves are lowered.

Interglacial. Period between two major advances of glacial ice. There were as many as eighteen expansions of glacial ice during the PLEISTOCENE ice age EPOCH. Scientists usually identify four major GLACIATIONS, with intervening interglacials. The names for the glacial stages are slightly different in Europe from those used in North America.

Interlocking spur. STREAM in a hilly or mountainous REGION that winds its way in a sinuous VALLEY between the different RIDGES, slowly eroding the ends of the spurs and straightening its course. The view of interlocking spurs looking upstream is a favorite of artists, as colors change with the receding distance of each interlocking spur.

Intermediate rock. IGNEOUS ROCK that is transitional between a basic and a silicic ROCK, having a SILICA content between 54 and 64 percent.

Intermittent lake. LAKE that is sometimes dry. See also PERENNIAL LAKE.

Intermittent stream. RIVER that has periods when its flow stops. See also PERENNIAL STREAM.

Internal drainage. Flow of a RIVER into an internal LAKE or SWAMP, rather than out to the SEA. If the lake has no outlet, SALTS accumulate over time, making the lake saline. This feature is called a salina. The Great Salt Lake is an example of this type of an internal DRAINAGE BASIN. If EVAPORATION is high, a dry salt LAKEBED is eventually produced. DESERT drainage is usually internal, with EPHEMERAL STREAMS and RUNOFF draining downward to the lowest part of a depression. Also known as interior drainage.

Internal migration. Movement of people within a COUNTRY, from one REGION to another. Internal MIGRATION in high-income economies is often urban-to-RURAL, such as the migration to the SUN BELT in the United States. In low-income economies, rural-to-URBAN migration is more common.

International date line. Line in the Pacific Ocean where each new DAY begins as the earth rotates. Most of the line is on the MERIDIAN at 180 DEGREES west (also east) LONGITUDE, but some irregularities occur to accommodate the wishes of individual ISLANDS.

International migration. Movement of people across an international

BOUNDARY, usually on a permanent basis. The source REGIONS for international migration to the United States changed from Northern Europe to Southern and Central Europe in the nineteenth and early twentieth century, and to LATIN AMERICA, with an increasing component from Asian countries, in the late twentieth century. In 1998, the United Nations estimated that more than 100 million people lived outside their COUNTRY of origin. Although many were refugees or political migrants seeking asylum, most were economic migrants seeking a better life and higher standard of living.

Intertillage. Mixed planting of different seeds and seedling crops within the same SWIDDEN or cleared patch of agricultural land. Potatoes, yams, corn, rice, and bananas might all be planted. The planting times are staggered throughout the year to increase the variety of crops or nutritional balance available to the subsistence farmer and his or her family.

Intertropical convergence zone (ITCZ). Line at which WINDS converge near the EQUATOR, because constant high INSOLATION and twelve hours of daylight cause AIR in this REGION to heat and rise. The rising air expands and cools, producing a band of CLOUDS and frequent PRECIPITATION, often in the form of THUNDERSTORMS. The ITCZ corresponds to the THERMAL EQUATOR.

Intrusive rock. IGNEOUS ROCK which was formed from a MAGMA that cooled below the surface of the earth; it is commonly coarse-grained.

Inversion. See TEMPERATURE INVERSION.

Ionosphere. Layer of the earth's ATMOSPHERE in which there are a large number of ions, or electrically charged particles, chiefly nitrogen and oxygen. The ionosphere begins at a height of about 30 miles (50 km.) above the earth's surface and extends up to about 240 miles (400 km.), but it is most distinct at ALTITUDES above about 50 miles (80 km.). The ionosphere contains three distinct layers—the D layer, E layer, and F layer. These layers are important to radio broadcasts, because they re-

Interaction between solar wind and the earth's ionosphere produces glowing light effects known as the Aurora borealis in the Northern Hemisphere and Aurora australis in the Southern Hemisphere. (PhotoDisc)

flect short-wave and AM radio transmission waves, especially at night; during the DAY, INSOLATION interferes with transmission. Ham radio operators use the bands of the ionosphere to communicate from their home base to distant parts of the earth. Television and FM signals are not affected by the ionosphere. The interaction of the SOLAR WIND with the earth's ionosphere produces glowing light effects known as the AURORA borealis and aurora australis.

Irredentism. Expansion of one COUNTRY into the territory of a nearby country, based on the residence of nationals in the neighboring country. Hitler used irredentist claims to invade Czechoslovakia, because small groups of German-speakers lived there in the Sudetenland. The term comes from Italian, referring to Italy's claims before World War I that all Italian-speaking territory should become part of Italy.

Irrigation. Bringing of water into drier REGIONS in order to use it for AGRICULTURE. Regions that have low RAINFALL or a long dry season use irrigation to ensure crop success. Modern TECHNOLOGY has enabled the construction of huge DAMS that produce HYDROELECTRIC POWER and also deliver water for irrigation by pipelines or CANALS. Surface irrigation includes flooding entire fields and furrow irrigation—running water between individual rows of plants. Alternatives are sprinkler irrigation systems—either an automatic traveling sprinkler system, in which a trailer moves a long arm of sprinklers slowly across a whole field, or a center-point pivot sprinkler that sprays a huge circular area—and drip irrigation, which delivers small amounts of water to each plant, using less water than other forms of irrigation. Water losses

Modern technology has made possible more efficient systems of irrigation, such as sprinklers, and with them, greater agricultural productivity. (PhotoDisc)

Furrow irrigation runs water between rows of plants. (PhotoDisc)

through EVAPORATION are a major concern with irrigation. The use of GROUNDWATER for irrigation has led to serious depletion of AQUIFERS worldwide. The Ogallala Aquifer in the United States lost more than 60 percent of its volume in the last third of the twentieth century and is not being replenished. Another major problem associated with irrigated agriculture is SALINIZATION.

Isallobar. Imaginary line on a MAP or meteorological chart joining PLACES with an equal change in ATMOSPHERIC PRESSURE over a certain time, often three hours. Isallobars indicate a pressure tendency and are used in WEATHER FORECASTING.

Islam. Religious faith with the second-largest number of adherents in the world, after Christianity. Its members are called Muslims. The word "Islam" means submission, obedience to the will of God. Islam recognizes the Old Testament prophets of the Bible but also believes that Muhammad (Mohammed) was the last of the prophets who brought God's words to earth. Muhammad was born in Medina, in what is now Saudi Arabia, in the seventh century. The holy book of Islam is the Qur'an (Koran). The two major branches of the Islamic faith are Sunni and Shia (Shiite).

Island. Piece of land, smaller than a CONTINENT, that is surrounded entirely by water. The world's largest island is Greenland. Islands are divided into four major types, depending on their formation: continental, oceanic, coral, and BARRIER. The isolation of islands has led to many interesting adaptations, and the study of island ECOSYSTEMS has been an exciting area of geography and biology. Charles Darwin's study of variations in finches on the Galapagos Islands was the foundation of his theories on evolution and NATURAL SELECTION. See also ARCHIPELAGO.

Island arc. Chain of VOLCANOES next to an oceanic TRENCH in the OCEAN BASINS; an oceanic PLATE descends, or subducts, below another oceanic plate at ISLAND arcs.

Islet. Small ISLAND.

Isobar. Imaginary line joining PLACES of equal ATMOSPHERIC PRESSURE. WEATHER MAPS show isobars encircling areas of high or low pressure. The spacing between isobars is related to the pressure gradient.

Isobath. Line on a MAP or CHART joining all PLACES where the water depth is the same; a kind of underwater CONTOUR LINE. This kind of map is a BATHYMETRIC CONTOUR.

Isoclinal folding. When the earth's CRUST is folded, the size and shape of the folds vary according to the force of compression and nature of the ROCKS. When the surface is compressed evenly so that the two sides of the fold are parallel, isoclinal folding results. When the sides or slopes of the fold are unequal or dissimilar in shape and angle, this can be an asymmetrical or overturned fold. See also ANTICLINE; SYNCLINE.

Isogonic line. Imaginary line drawn on a MAP connecting PLACES that have the same deviation from true north when a magnetic needle or COMPASS is used. This is necessary because the earth's NORTH MAGNETIC POLE does not correspond with the North Pole of 90 DEGREES north that represents true north or GRID north. Since the earth's magnetic north is not in a fixed position, isogonic lines vary over time. Aircraft pilots make use of isogonic lines.

Isohaline. Imaginary line on a CHART connecting points of equal SALINITY.

Isohel. Imaginary line drawn on a MAP connecting PLACES that receive an equal amount of sunshine.

Isohyet. Imaginary line drawn on a MAP connecting PLACES with the same amount of PRECIPITATION over a given time. Average annual RAINFALL is shown with isohyets. Seasonal precipitation maps are commonly constructed using isohyets. As a general guide, the isohyet marking 10 inches (250 millimeters) of annual precipitation is the lower limit beyond which crops cannot be grown without IRRIGATION.

Isoline. Imaginary line drawn on a MAP along which there is a constant value of the factor under study. Isolines commonly used by geographers include ISOTHERMS, ISOBARS, ISOHYETS, and CONTOUR LINES.

Isomagnetic charts. MAPS on which are traced curves, all the points of which have the same value in some magnetic element.

Isopleth. Imaginary line drawn on a MAP connecting points of equal value, based on calculations of various climatic variables such as average daily TEMPERATURE, average monthly PRECIPITATION, or number of FROST DAYS. An important isopleth to foresters is the TIMBERLINE, which marks an elevation above which it becomes too cold for trees to grow. In the KOEPPEN CLIMATE CLASSIFICATION, this is the BOREAL FOREST-TUNDRA BOUNDARY, which is calculated as PLACES where at least one month a year has an average temperature of at least 50 DEGREES Fahrenheit (10 degrees Celsius).

Isoseismal line. Line constructed after an EARTHQUAKE, showing areas of equal intensity of the earthquake. The intensity is calculated using seismographic records, along with study of the effects on buildings and surfaces. Intensity is a MEASUREMENT that combines data regarding ground shaking, features of the SEISMIC WAVES, geology, and other factors. If the earth's CRUST were completely uniform, the isoseismal curves would be concentric circles surrounding the EPICENTER of the earthquake, with the highest intensity at the center. Other factors, such as ROCK properties and positions of FAULTS, can cause an asymmetric pattern of isoseismal lines; at times, the epicenter is not located in the area of highest intensity. A modified MERCALLI SCALE is used in the United States to record earthquake intensity. It uses the numbers I (not felt) through XII (nearly total damage) to describe increasing intensity. Intensity is not the same as MAGNITUDE, which is commonly used to describe the size of an earthquake numerically.

Isostasy. Theory that the earth's CRUST maintains equilibrium because it is supported on a denser layer, at a depth of about 60 miles (100 km.). When LANDFORMS are weathered and eroded, material is removed from one PLACE, such as a mountain RANGE, and deposited in another, such as a FLOODPLAIN. On a large scale, material is eroded from CONTINENTS and deposited on OCEAN floors. Because of isostasy, however, the continents do not completely wear down. In response to the decrease in weight, the earth's crust rises. This process is called isostatic readjustment. It is similar to the fact that unloading a ship causes it to rise higher in the water. The removal of the great weight of glacial ice over the continents during the last ICE AGE led to isostatic UPLIFT, which is especially marked in Scandinavia. There, the rate of uplift is calculated at approximately 0.4 inch (1 centimeter) per year over the last few centuries.

Isotherm. Line joining PLACES of equal TEMPERATURE. A world MAP with isotherms of average monthly temperature shows that over the OCEANS, temperature decreases uniformly from the EQUATOR to the POLES, and higher temperatures occur over the CONTINENTS in summer and lower temperatures in winter because of the unequal heating properties of land and water.

Isotropic. Having properties the same in all directions; if elastic waves propagate at the same velocity in all directions, they are isotropic.

Isotropic surface. Hypothetical flat surface or PLAIN, with no variation in any physical attribute. An isotropic surface has uniform ELEVATION, SOIL type, CLIMATE, and VEGETATION. Economic geographic models study behavior on an isotropic surface before applying the results to the real world. For example, in an isotropic model, land value is highest at the CITY center and falls regularly with increasing distance from there. In the real world, land values are affected by elevation, water features, URBAN regulations, and other factors. The von Thuenen model of the Isolated State is based on a uniform plain or isotropic surface.

Isthmian links. Chains of ISLANDS between substantial LANDMASSES.

Isthmus. Narrow strip of land connecting two larger bodies of land. The Isthmus of Panama connects North and South America; the Isthmus of Suez connects Africa and Asia. Both of these have been cut by CANALS to shorten shipping routes.

ITCZ. See INTERTROPICAL CONVERGENCE ZONE.

Jebel. Arabic word for MOUNTAIN.

Jet stream. WINDS that move from west to east in the upper ATMOSPHERE, 23,000 to 33,000 feet (7,000-10,000 meters) above the earth, at about 200 miles (300 km.) per hour. They are narrow bands, elliptical in cross section, traveling in irregular paths. Four jet streams of interest to earth scientists and meteorologists are the polar jet stream and the subtropical jet stream in the Northern and SOUTHERN HEMISPHERES. The polar jet stream is located at the TROPOPAUSE, the BOUNDARY be-

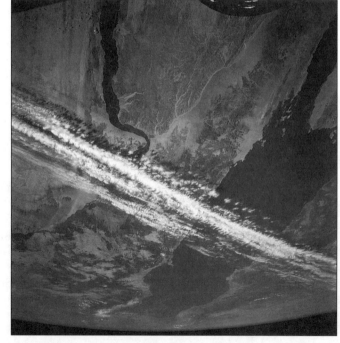

Jet stream passing over northern Egypt and the Red Sea at a speed of about one hundred miles per hour. The Nile River can be seen on the left and the southern tip of the Sinai Peninsula at the upper right. (Corbis)

tween the TROPOSPHERE and the STRATOSPHERE, along the polar FRONT. There is a complex interaction between surface winds and jet streams. In winter the NORTHERN HEMISPHERE polar front can move as far south as Texas, bringing BLIZZARDS and extreme WEATHER conditions. In summer, the polar jet stream is located over Canada. The subtropical jet stream is located at the tropopause around 30 DEGREES north or south LATITUDE, but it also migrates north or south, depending on the SEASON. At times, the polar and subtropical jet streams merge for a few DAYS. Aircraft take advantage of the jet stream, or avoid it, depending on the direction of their flight. Upper atmosphere winds are also known as GEOSTROPHIC winds.

Jetty. Structure built to protect a HARBOR entrance from WAVE EROSION or to prevent DEPOSITION.

Joint. Naturally occurring fine crack in a ROCK, formed by cooling or by other stresses. SEDIMENTARY ROCKS can split along bedding planes; other joints form at right angles to the STRATA, running vertically through the rocks. In IGNEOUS ROCKS such as GRANITE, the stresses of cooling and contraction cause three sets of joints, two vertical and one parallel to the surface, which leads to the formation of distinctive LANDFORMS such as TORS. BASALT often demonstrates columnar jointing, producing tall columns that are mostly hexagonal in section. The presence of joints in BEDROCK hastens WEATHERING, because water can penetrate into the joints. This is particularly obvious in LIMESTONE, where joints are rapidly enlarged by SOLUTION. FROST WEDGING is a type of PHYSICAL WEATHERING that can split large boulders through the expansion when water in a joint freezes to form ice. Compare with FAULTS, which occur through tectonic activity.

Frost-split granite boulder. (U.S. Geological Survey)

Jungle. Degenerate form of tropical RAIN FOREST that grows where the upper, closed-tree canopy is absent, allowing smaller trees and shrubs to flourish. In this dense, leafy VEGETATION, a machete is needed to hack a path through the luxuriant plant growth. In a true tropical rain forest, the upper closed canopy of leaves prevents sunlight from reaching the forest floor, so there is little undergrowth, and a person can walk through easily. Jungle occurs naturally along the BANKS of RIVERS or when a STORM fells trees of the forest. When an area of rain forest is cleared for subsistence farming and later abandoned, jungle is an intermediate stage in the return to true rain forest. However, POPULATION pressure in many countries means that increasing areas of tropical rain forest cannot regenerate.

Jurassic. Second of the three PERIODS that make up the MESOZOIC ERA. It occurred around 208 to 144 million years ago and lasted for around 64 million years. Dinosaurs lived on Earth during the Jurassic, including the giant vegetarians and the smaller CARNIVORES. Birds and small mammals appeared during this time; ammonites, sharks, plesiosaurs, and bony fish lived in the SEAS. The dominant plants were cycads, together with CONIFEROUS FORESTS. The name comes from the Jura Mountains of France and Switzerland.

Kame. Small HILL of gravel or mixed-size deposits, SAND, and gravel. Kames are found in areas previously covered by CONTINENTAL GLACIERS or ICE SHEETS, near what was the outer edge of the ice. They may have formed by materials dropping out of the melting ice, or in a deltalike deposit by a STREAM of MELTWATER. These deposits of which kames are made are called drift. Small LAKES called KETTLES are often found nearby. A closely spaced group of kames is called a kame field.

Kames in Happy Valley in Greenland's Nunatarssuaq region. (U.S. Geological Survey)

Unusual karst formations in Utah's Goblin Valley. (Corbis)

Karst. LANDSCAPE of SINKHOLES, underground STREAMS and caverns, and associated features created by CHEMICAL WEATHERING, especially SOLUTION, in REGIONS where the BEDROCK is LIMESTONE. The name comes from a region in the southwest of what is now Slovenia, the Krs (Kras) Plateau, but the karst region extends south through the Dinaric Alps bordering the Adriatic Sea, into Bosnia-Herzegovina and Montenegro. Where limestone is well jointed, RAINFALL penetrates the JOINTS and enters the GROUNDWATER, carrying the MINERALS, especially calcium, away in solution. Most of the famous CAVES and caverns of the world are found in karst areas. The Carlsbad Caverns in New Mexico are a good example. Kentucky, Tennessee, and Florida also have well-known areas of karst. In some tropical countries, a form called tower karst is found. Tall conical or steep-sided HILLS of limestone rise above the flat surrounding landscape. Around 15 percent of the earth's land surface is karst TOPOGRAPHY.

Katabatic wind. GRAVITY DRAINAGE WINDS similar to MOUNTAIN BREEZES but stronger in force and over a larger area than a single VALLEY. Cold AIR collects over an elevated REGION, and the dense cold air flows strongly downslope. The ICESHEETS of Antarctica and Greenland produce fierce katabatic winds, but they can occur in smaller regions. The BORA is a strong, cold, squally downslope wind on the Dalmatian COAST of Yugoslavia in winter.

Kettle near Tuolumne Meadows in California's Yosemite National Park. (U.S. Geological Survey)

Kettle. Small depression, often a small LAKE, produced as a result of continental GLACIATION. It is formed by an isolated block of ice remaining in the ground MORAINE after a GLACIER has retreated. Deposited material accumulates around the ice, and when it finally melts, a steep hole remains, which often fills with water. Walden Pond, made famous by writer Henry David Thoreau (1817-1862), is a glacial kettle.

Key. Small coral ISLAND; a sandy island built up by WAVE action on a CORAL REEF. The Florida Keys are a good example from the United States. Called CAYS in other countries.

Khamsin. Hot, dry, DUST-laden WIND that blows in the eastern Sahara, in Egypt, and in Saudi Arabia, bringing high TEMPERATURES for three or four DAYS. Winds can reach GALE force in intensity. The word Khamsin is Arabic for "fifty" and refers to the period between March and June when the khamsin can occur.

Knickpoint. Abrupt change in gradient of the bed of a RIVER or STREAM; sometimes spelled nickpoint. It is marked by a WATERFALL, which over time is eroded by FLUVIAL action, restoring the smooth profile of the riverbed. The knickpoint acts as a TEMPORARY BASE LEVEL for the upper part of the stream. Knickpoints can occur where a hard layer of ROCK is slower to erode than the rocks downstream, for example at Niagara Falls. Other knickpoints and waterfalls can develop as a result of tectonic forces. UPLIFT leads to new EROSION by a stream, creating a knickpoint that gradually moves upstream. The bed of a tributary GLACIER is often considerably higher than the VALLEY of the main glacier,

Yosemite Falls, the highest waterfall in Yosemite National Park, is fed by melting glacial ice at a higher altitude. (PhotoDisc)

so that after the glaciers have melted, a waterfall emerges over this knickpoint from the smaller hanging valley to join the main stream. Yosemite National Park has several such waterfalls.

Koeppen climate classification. Commonly used scheme of CLIMATE classification that uses statistics of average monthly TEMPERATURE, average monthly PRECIPITATION, and total annual precipitation. The system was devised by Wladimir Koeppen early in the twentieth century.

Kopje. South African word for a small flat-topped LANDFORM; called a BUTTE in the United States.

La Niña. WEATHER phenomenon that is the opposite part of EL NIÑO. When the SURFACE WATER in the eastern Pacific Ocean is cooler than average, the southeast TRADE WINDS blow strongly, bringing heavy rains to countries of the western Pacific. Scientists refer to the whole RANGE of TEMPERATURE, pressure, WIND, and SEA LEVEL changes as the SOUTHERN OSCILLATION (ENSO). The term "El Niño" gained wide currency in the American media after a strong ENSO warm event in 1997-1998. A weak ENSO cold event, or La Niña, followed it in 1998. Means "the little girl" in Spanish. Alternative terms are "El Viejo" and "anti-El Niño."

Laccolith. LANDFORM of INTRUSIVE volcanism formed when viscous MAGMA is forced between overlying sedimentary STRATA, causing the surface to bulge upward in a domelike shape.

Lagoon. Area of shallow, quiet water, separated from the OCEAN by a natural barrier. There are two types of lagoons—coastal and coral. Coastal lagoons are long and narrow, separated from the SEA by a SANDBAR, with a narrow outlet to the sea. The water height in the lagoon changes with the TIDE. Over time the continued supply of SEDIMENT by STREAMS may lead to the infilling of a coastal lagoon, so that it becomes a WETLAND or SALT MARSH. A coral lagoon is found where a barrier REEF separates the land from the ocean. In the case of an ATOLL, there is no land, only the ring of coral surrounding the lagoon.

Lagoon in Kauai in the Hawaiian Islands. (PhotoDisc)

Lahar. Type of mass movement in which a MUDFLOW occurs because of a volcanic explosion or ERUPTION. The usual cause is that the heat from the LAVA or other pyroclastic material melts ice and SNOW at the VOLCANO's SUMMIT, causing a hot mudflow that can move downslope with great speed. The eruption of Mount Saint Helens in 1985 was accompanied by a lahar.

One of the largest artificial lakes in the world, Lake Mead (pictured in 1985) was created by the construction of Hoover Dam, which traps the Colorado River, as it enters from the east (to the right), and other tributaries. Creation of the lake along the Nevada-Arizona border has helped make possible the growth of nearby Las Vegas (left) and the national recreation area that surrounds the irregularly shaped lake. (Corbis)

Lake. Large body of water enclosed in a BASIN. STREAMS enter and leave it, so there is a slow movement of water through the lake. If a lake has no outlet, its water becomes saline. There is considerable confusion over terminology because the world's largest lake is saline and is called the Caspian Sea; however, it is not a true SEA, nor is its neighbor, the Aral Sea. The world's largest freshwater lake in terms of surface area is North America's Lake Superior. The lake with the greatest volume of FRESH WATER is Lake Baikal in Siberia. There also are artificial lakes created by human activities such as DAM construction. Lake Powell on the Colorado River is a controversial example.

Lake basin. Enclosed depression on the surface of the land in which SUR-FACE WATERS collect; BASINS are created primarily by glacial activity and tectonic movement.

Lakebed. Floor of a LAKE.

Land breeze. Local WIND that is the opposite of a sea breeze. During the evening, when the land near the COAST cools more rapidly than the adjacent OCEAN, AIR rises above the warmer water, forming a low-pressure REGION. A BREEZE develops as air from over the land moves toward this lower pressure.

Land bridge. Piece of land connecting two CONTINENTS, which permits the MIGRATION of humans, animals, or plants from one area to another. Many former land bridges are now under water, because of the rise in SEA LEVEL after the last ICE AGE. The Bering Strait connecting Asia and North America was an important land bridge for the latter continent.

Land hemisphere. Because the DISTRIBUTION of land and water surfaces on Earth is quite asymmetrical on either side of the EQUATOR, the NORTHERN HEMISPHERE might well be called the land hemisphere. For many centuries, Europeans refused to believe that there was not an equal area of land in the SOUTHERN HEMISPHERE. Explorers such as James Cook were dispatched to seek such a "Great South Land."

Land use. Predominant activity over an area. Common land uses include AGRICULTURE, forestry, NATIONAL PARKS, and reserves. In modern times, more land is being used for URBAN residential and industrial purposes and for roads and other INFRASTRUCTURE. Often, the best agricultural land is thereby taken out of production.

Landform. Conspicuous feature on the surface or CRUST of the earth, including underwater. Landforms are also studied on other PLANETS, using vehicles such as the Mars Explorer. Common landforms include MOUNTAINS of various kinds, PLATEAUS, and VALLEYS. Landforms can be understood through a study of the processes that led to their formation. Tectonic landforms are produced through CRUSTAL MOVEMENT. Volcanic landforms result from cooling of MAGMA. Structural landforms are produced by EROSION or DEPOSITION by the forces of STREAMS, GLACIERS, WAVES, and WINDS. Biogenic landforms are produced by organisms such as coral or termites, or, more extensively, by humans. GEOMORPHOLOGY is the study of the origin and development of landforms.

Landlocked country. NATION that is surrounded by other countries and does not have an OCEAN COAST. This is an economic disadvantage in terms of PORT facilities and international TRADE, since exports and imports must pass through a neighboring COUNTRY. A landlocked country also has no control of the fishing and OIL RESOURCES of the CONTINENTAL SHELF or an exclusive economic zone, both of which are enjoyed by countries with a COASTLINE. Bolivia and Paraguay are landlocked countries in South America. Many countries in Asia, Europe, and Africa are landlocked.

Landmass. Large area of land—an ISLAND or a CONTINENT.

Landsat. Space-exploration project begun in 1972 to MAP the earth continuously with SATELLITE imaging. The satellites have collected data about the earth: its AGRICULTURE, FORESTS, flat lands, MINERALS, waters, and ENVIRONMENT. These were the first satellites to aid in Earth sciences, helping to produce the best maps available and assisting farmers around the world to improve their crop yields.

Landscape. Natural landscape is made up of LANDFORMS that reflect the processes operating in the area for greater or shorter periods of time. A KARST landscape, for example, consists of distinctive landforms such as SINKHOLES, UVALA, DOLINES, and caverns. A glacial landscape shows the results of EROSION or DEPOSITION by ice. Most landscapes are shaped by FLUVIAL processes, or STREAMS and running water.

Landslide. Sudden, rapid downslope movement of earth or ROCK, although the latter is also called a rockslide; one of the forms of mass movement. In a landslide, a section of the hillside moves as a cohesive mass along a plane parallel to the slope angle. Landslides can be caused by undercutting at the base, for example in road construction or in excavation to create a building SITE. A landslide that has a rotational component to its movement is called a SLUMP. The explosion of Mount Saint Helens involved a huge landslide triggered by an EARTHQUAKE prior to the explosion.

Language. Means of human communication. It is estimated that there are more than six thousand languages in use in the world today. Some exist only as spoken languages, but most have a written form, using symbols to record the language. There are also extinct languages that exist only as written records, such as Ancient Egyptian hieroglyphics, although some extinct languages are used for religious purposes, such as Sanskrit in Buddhism. Linguists place English in the Indo-European LANGUAGE FAMILY.

Language branch. Collection of related LANGUAGES that have developed from a common ancestor but have experienced changes over time, leading to variations in language. English belongs to the Germanic language branch of the Indo-European LANGUAGE FAMILY. Other languages in the Germanic language branch include Dutch, Swedish, Danish, Norwegian, and Icelandic. Another language branch of the Indo-European family is the Romance branch, which includes the French, Spanish, Portuguese, Italian, and Romanian languages.

Language family. Group of related LANGUAGES believed to have originated from a common prehistoric language. English belongs in the Indo-European language family, which includes the languages spoken by half of the world's peoples.

Lapilli. Small ROCK fragments that are ejected during volcanic ERUPTIONS. A lapillus ranges from about the size of a pea to not larger than a walnut. Some lapilli form by accretion of VOLCANIC ASH around moisture droplets, in a manner similar to hailstone formation. Lapilli sometimes form into a textured rock called lapillistone.

Late Precambrian era. That part of geologic time from about 550 million years to 1 billion years before the present.

Laterite. Bright red CLAY SOIL, rich in iron oxide, that forms in tropical CLIMATES, where both TEMPERATURE and PRECIPITATION are high year-round, as ROCKS weather. It can be used in brick making and is a source of iron. When the soil is rich in aluminum, it is called BAUXITE. When laterite or bauxite forms a hard layer at the surface, it is called duricrust. Australia and sub-Saharan Africa have large areas of duricrust, some of which is thought to have formed under previous conditions during the TRIASSIC period.

Latin America. WESTERN HEMISPHERE REGION generally regarded as including Mexico, Central America, most of the islands of the Caribbean, and the entire continent of South America. After the voyages of Christopher Columbus, the peoples of Latin America were conquered and colonized by the Spaniards and Portuguese, starting in the late fifteenth century.

Latitude. Measure of distance north or south on the earth's surface. Lines of latitude (also called PARALLELS) are imaginary lines running east-west around the globe. They are numbered from zero DEGREES at the EQUATOR to ninety degrees north or south at the North or South Pole, respectively. Each degree of latitude, measured along a MERIDIAN, is about 69 miles (111 km.) in length. The distance varies slightly because of the flattening of the earth towards the POLES. Lines of latitude decrease in length from 24,902 miles (40,075 km.) at the equator to a single point at the poles. The most important lines of latitude are the equator, at 0 degrees; tropic of CANCER, 23.5 degrees north; Arctic Circle, 66.5 degrees north; North Pole, 90 degrees north; tropic of CAPRICORN, 23.5 degrees south, Antarctic Circle, 66.5 degrees south; and South Pole, 90 degrees south. The equator is the only line of latitude that is a GREAT CIRCLE.

Laurasia. Hypothetical SUPERCONTINENT made up of approximately the present CONTINENTS of the NORTHERN HEMISPHERE.

Lava. MAGMA, or molten material from within the earth, that emerges at

Lava flow in Hawaii. (PhotoDisc)

the surface. It forms EXTRUSIVE IGNEOUS ROCKS such as BASALT and obsidian. The Hawaiians distinguish between two types of lava: Pahoehoe is smooth, fluid, flowing lava that hardens into ripples and folds; AA is rough, broken, jagged pieces of rock.

Lava tube. Cavern structure formed by the draining out of liquid LAVA in a pahoehoe flow.

Layered plains. Smooth, flat REGIONS believed to be composed of materials other than sulfur compounds.

Leaching. Removal of nutrients from the upper horizon or layer of a SOIL, especially in the humid TROPICS, because of heavy RAINFALL. The remaining soil is often bright red in color because iron is left behind. Despite their bright color, tropical soils are infertile.

Leeward. Rear or protected side of a MOUNTAIN or RANGE is the leeward side. Compare to WINDWARD.

Legend. Explanation of the different colors and symbols used on a MAP. For example, a map of the world might use different colors for high-income, middle-income, and low-income economies. A historical map might use different colors for countries that were once colonies of Britain, France, or Spain.

Legumes. Type of plant in which the fruit is released by the splitting open of the fruit along two sides or seams. Legumes important to humans include peas, clover, alfalfa, beans of many kinds, and peanuts. High in protein, legumes are an important part of the human food supply. Bacteria that live in the roots of most legume crops fix nitrogen in the SOIL, so legumes are grown as part of CROP ROTATION, in order to restore soil fertility naturally, without the addition of FERTILIZERS.

Levee. NATURAL LEVEES are long, low RIDGES of ALLUVIUM formed at the RIVER BANK of STREAMS flowing on FLOODPLAINS. After the stream overflows onto the floodplain annually, the water velocity decreases sharply; material deposited there forms the levee. Behind levees are low-lying areas called backswamps. Humans have built artificial levees of earth, ROCK, or concrete to try to prevent water spreading onto the floodplain. The Mississippi River has the world's largest system of artificial levees. Rivers that transport large amounts of SEDIMENT generally deposit some of it in their bed during low flow, so over time the bed of the stream becomes higher than the surrounding floodplain. This leads to catastrophic flooding when a levee is breached. Both the Mississippi and the Huang He (Yellow), in China, are good examples of this happening.

Life expectancy. Average number of years that a newly born human can expect to live in any given society or COUNTRY. In the 1990's, life expectancy in high-income economies was more than seventy years, but in many low-income economies, especially in Africa, it was less than fifty years. Women had a life expectancy a few years higher than men in all countries.

Light year. Distance traveled by light in one year; widely used for measuring stellar distances, it is equal to roughly 6 trillion miles (9.5 million km.).

Lightning. Visible discharge of electric ENERGY in the earth's ATMOSPHERE; a giant electric arc passing from the CLOUD to the ground. Usually part of the activity associated with the growth of a CUMULONIMBUS cloud or thunderhead. A positive charge builds in the upper part of the cloud and a negative charge in the lower part. A flash of cloud-to-ground lightning involves a smaller leader stroke, followed by a brilliant return stroke. Eight million lightning strikes can occur each DAY on Earth. A lightning flash involves hundreds of millions of volts, and associated TEMPERATURES are as high as 54,000 DEGREES Fahrenheit (30,000 degrees Celsius). The heated AIR moving at supersonic speed causes the thunder that accompanies lightning. Metallic lightning rods attached to buildings attract lightning strikes and conduct the charge harmlessly to the ground.

Lightning storm. (PhotoDisc)

Lignite. Low-grade COAL, often called brown coal. It is mined and used extensively in eastern Germany, Slovakia, and the Moscow Basin.

Limestone. SEDIMENTARY ROCK comprising mainly calcium carbonate. Limestone is rich in FOSSIL remains, and their study has contributed

Some of the most unusual limestone formations in the world are found in Western Australia's Pinnacles Desert, in Nambung National Park. The limestone in these pillars—some of which are as much as ten feet tall—originated in ancient marine sea shell material brought ashore by waves and carried inland by wind. (Corbis)

greatly to our knowledge of Earth history. Distinctive LANDFORMS known as KARST are produced in areas of limestone.

Lingua franca. Latin for "language of the Franks," a mixed LANGUAGE used in the Roman Empire for TRADE. In modern usage, a widely understood and commonly spoken second language for many people. Globalization is leading to English becoming a lingua franca almost everywhere.

Liquefaction. Loss in cohesiveness of water-saturated SOIL as a result of ground shaking caused by an EARTHQUAKE.

Literacy rate. DEMOGRAPHIC MEASURE of what percentage of the adult POPULATION of a COUNTRY or REGION can read and write. Low-income countries have low literacy rates; less than 40 percent of adults are literate in countries such as Pakistan and Ethiopia, for example. Additionally, in such countries the literacy rate for women is considerably lower, as men are given access to education in greater numbers. In high-income countries, the literacy rate approaches 100 percent.

Lithic. Having to do with ROCK.

Lithification. Process whereby loose material is transformed into solid ROCK by compaction or cementation.

Lithology. Description of ROCKS, such as rock type, MINERAL makeup, and fluid in rock pores.

Lithosphere. Solid outermost layer of the earth. It varies in thickness from a few miles to more than 120 miles (200 km.). It is broken into pieces known as TECTONIC PLATES, some of which are extremely large, while others are quite small. The upper layer of the lithosphere is the CRUST, which may be CONTINENTAL CRUST or OCEANIC CRUST. Below the crust is a layer called the ASTHENOSPHERE, which is weaker and plastic, enabling the motion of tectonic plates.

Lithospheric crust. Relatively thin outer portion of Earth's "onion" structure, composed of solid ROCK.

Lithospheric plate. One of a number of crustal PLATES of various sizes that compose the earth's outer CRUST; their BORDERS are outlined by major zones of EARTHQUAKE activity.

Littoral. Adjacent to or related to a SEA.

Littoral current. See LONGSHORE CURRENT.

Livestock. Domesticated animals raised on farms or in agricultural communities for food. The term is usually applied to mammals such as sheep, pigs, cattle, and horses.

Llanos. Grassy REGION in the Orinoco Basin of Venezuela and part of Colombia. SAVANNA VEGETATION gradually gives way to scrub at the outer edges of the *llanos.* The area is relatively undeveloped.

Loam. SOIL TEXTURE classification, indicating a soil that is approximately equal parts of SAND, SILT, and CLAY. Farmers generally consider a sandy loam to be the best soil texture because of its water-retaining qualities and the ease with which it can be cultivated.

Local sea-level change. Change in SEA LEVEL only in one area of the world, usually by land rising or sinking in that specific area.

Local winds. WINDS that, over a small area, differ from the general pressure pattern owing to local thermal or orographic effects.

Location. Geographers identify two kinds of location—absolute and relative. ABSOLUTE LOCATION is a position given with coordinates of the geographic GRID, such as 33 DEGREES north and 118 degrees west. There is one precise spot on Earth corresponding to that absolute location. RELATIVE LOCATION is a verbal description of a PLACE with reference to some other place, for example, the "Middle East," the "Midwest," "Dixie."

Loch. Scottish term for a LAKE. Many lochs are products of EROSION by continental GLACIATION, and have narrow, elongated shapes. Scottish lochs are located in the Great Glen, which extends across the COUNTRY for almost 60 miles (100 km.), from Moray Firth to Loch Linnhe. The lakes are connected by the Caledonian Canal, built in the early nineteenth century by Thomas Telford. Loch Ness, in northern Scotland, is the largest freshwater lake in Great Britain. Loch Lomond, another large Scottish lake, located near Glasgow, is the subject of a well-known Scottish song.

Lode deposit. Primary deposit, generally a VEIN, formed by the filling of a FISSURE with MINERALS precipitated from a HYDROTHERMAL solution.

Loess. EOLIAN, or wind-blown, deposit of fine, silt-sized, light-colored material. Loess covers about 10 percent of the earth's land surface. The loess PLATEAU of China is good agricultural land, although susceptible to EROSION. Loess has the property of being able to form vertical CLIFFS or BLUFFS, and many people have built dwellings in the steep cliffs above the Huang He (Yellow) River. In the United States, loess deposits are found in the VALLEYS of the Platte, Missouri, Mississippi, and Ohio Rivers, and on the Columbian Plateau. A German word, meaning loose or unconsolidated, which comes from loess deposits along the Rhine River.

Longitude. Measure of angular distance on the earth's surface, east or west of the PRIME MERIDIAN. Lines of longitude (called MERIDIANS) are imaginary lines, numbered from 0 DEGREES at the prime meridian through 180 degrees, either east or west. They converge at the North and South Poles. One degree of longitude is 69 miles (111 km.) at the EQUATOR, but less than half of that distance at 60 degrees north or south LATITUDE, reducing to zero at the POLES. Each meridian is half of a GREAT CIRCLE.

Longitudinal bar. Midchannel accumulation of SAND and gravel with its long end oriented roughly parallel to the RIVER flow.

Longitudinal dune. Elongate SAND DUNE parallel to the prevailing WIND.

Longshore current. Current in the OCEAN close to the SHORE, in the surf zone, produced by WAVES approaching the COAST at an angle. Also called a LITTORAL current. The longshore current combined with wave action can move large amounts of SAND and other BEACH materials down the coast, a process called LONGSHORE DRIFT.

Longshore drift. The movement of SEDIMENT parallel to the BEACH by a LONGSHORE CURRENT.

Low island. ISLAND made of coral and coral SAND, especially common in the Pacific Ocean. Because these islands have low ELEVATION, they receive no OROGRAPHIC PRECIPITATION and AGRICULTURE is not possible. Therefore, low islands can support only small POPULATIONS. They are in danger of inundation as GLOBAL WARMING leads to a rise in SEA LEVEL.

Low velocity zone. See ASTHENOSPHERE.

Lunar eclipse. See ECLIPSE, LUNAR.

Maar. Explosion vent at the earth's surface where a volcanic cone has not formed. A small ring of pyroclastic materials surrounds the maar. Often a LAKE occupies the small CRATER of a maar. A larger form is called a TUFF RING.

Macroburst. Updrafts and downdrafts within a CUMULONIMBUS CLOUD or THUNDERSTORM can cause severe TURBULENCE. A DOWNBURST within a thunderstorm when windspeeds are greater than 130 miles (210 km.) per hour and over areas of 2.5 square miles (5 sq. km.) or more is called a macroburst. See also MICROBURST.

Macrofossil. FOSSIL large enough to study with the unaided eye, as opposed to a microfossil, which requires a microscope for examination.

Magma. Body of molten ROCK, including any dissolved gases and suspended crystals.

Magnetic declination. See DECLINATION, MAGNETIC.

Magnetic field. Magnetic lines of force that are projected from the earth's interior and out into space.

Magnetic poles. Locations on the earth's surface where the earth's MAGNETIC FIELD is perpendicular to the surface. The magnetic poles do not correspond exactly to the geographic North Pole and South Pole, or earth's AXIS; the difference is called magnetic variation or DECLINATION.

Magnetic reversal. Change in the earth's MAGNETIC FIELD from the North Pole to the South MAGNETIC POLE.

Magnetic storm. Rapid changes in the earth's MAGNETIC FIELD as a result of the bombardment of the earth by electrically charged particles from the SUN.

Magnetic survey. MEASUREMENTS of the magnetic elements at many points, on or above the earth's surface, carried out by field teams, airborne magnetometers, ships at SEA, or SATELLITES.

Magnetism. The MAGNETIC FIELD of Earth is like a bar magnet, with one end being the North Pole and the opposite end the South Pole. The MAGNETOSPHERE extends on average more than one hundred miles above the earth's surface. The principal source of the magnetism is the movement of the liquid OUTER CORE, which is heated by radioactive decay of the INNER CORE. The use of a magnet COMPASS for NAVIGATION was known in ancient times. Earth's magnetic field has reversed its POLARITY many times; studies of PALEOMAGNETISM at the mid-Atlantic Ridge were an important contribution to the development of the modern theory of PLATE TECTONICS. Magnetic currents extend into the ATMOSPHERE, protecting the earth from the SOLAR WIND, and also causing effects known as AURORAS.

Magnetosphere. REGION surrounding a PLANET where the planet's own MAGNETIC FIELD predominates over magnetic influences from the SUN or other planets.

Mandate. Term applied by the League of Nations to the German colonies it assigned to the administration of Great Britain, France, and South

Africa after World War I. After the United Nations succeeded the League of Nations in the 1940's, the mandate territories were officially redesignated trust territories.

Mangrove swamp. WETLAND, similar to a midlatitude SWAMP. Along low-lying COASTS in the TROPICS, and in some subtropical areas, coasts are forested in low halophytic trees called mangroves. These mangroves, and some associated plants, grow in tidal LAGOONS and estuaries in muddy, anaerobic conditions. Despite their impenetrable nature and their odiferous qualities, mangrove swamps form a highly productive ECOSYSTEM.

Mantle. Part of the earth below the CRUST, surrounding the CORE. The separation between the crust and the mantle is called the Mohorovičić discontinuity, shortened often to the "Moho." The mantle is approximately 1,800 miles (2,900 km.) thick, comprising more than 80 percent of Earth's volume but only two-thirds of its weight, since the core is much denser. Geophysicists differentiate between the UPPER MANTLE (about 600 miles/1,000 km. thick) and the lower mantle. The uppermost part of the mantle is the ASTHENOSPHERE.

Mantle convection. Thermally driven flow in the earth's MANTLE thought to be the driving force of PLATE TECTONICS.

Mantle plume. Rising jet of hot MANTLE material that produces tremendous volumes of basaltic LAVA. See also HOT SPOT.

Map. Representation of all or part of the earth's surface at a smaller size. A globe is the only accurate map of Earth, since transformation of a three-dimensional body to a two-dimensional surface such as a sheet of paper involves distortions of shape, size, and direction. A map used for NAVIGATION is referred to as a CHART. A map used for determining and recording property boundaries is a PLAT. The art and science of map-making is CARTOGRAPHY. See also AZIMUTHAL PROJECTION; CONICAL PROJECTION; CYLINDRICAL PROJECTION.

Map projection. Mathematical formula used to transform the curved surface of the earth onto a flat plane or sheet of paper. Projections are divided into three classes: CYLINDRICAL, CONICAL, and AZIMUTHAL.

Maquiladora. Term for modern factories and industrial establishments in Mexico, where foreign components are assembled into products for export, especially to the United States. Maquiladoras are usually owned by American or Japanese transnational companies and are usually located near the U.S. BORDER. Mexico benefits through increased employment and worker training; the creation of the North American Free Trade Agreement (NAFTA) meant a reduction of import duties, making the goods cheaper to American purchasers. Cities on the U.S.-Mexico border are economically favorable locations for maquiladoras, and Tijuana and Ciudad Juarez are the two largest.

Marble. LIMESTONE that has been crystallized by heat and pressure. The process of recrystallization destroys FOSSILS as they change into calcite.

The Renaissance artist Michelangelo carved his famous statue David from white marble in the early sixteenth century. The statue now stands in Florence, Italy. (PhotoDisc)

Marble comes in many colors, but sculptors such as Michelangelo historically preferred the pure white marble found at Cararra in Italy.

Marchland. FRONTIER area where boundaries are poorly defined or absent. The marches themselves were a type of BOUNDARY REGION. Marchlands have changed hands frequently throughout history. The name is related to the fact that armies marched across them.

Marine. Pertaining to a seawater, OCEAN ENVIRONMENT.

Market town. Small TOWN or VILLAGE that holds regular public markets. Historically, market towns were gathering points in larger areas.

Marl. Type of CLASTIC ROCK that is a naturally occurring mixture of CLAY and LIMESTONE. The fine-grained calcareous material can originate under freshwater conditions, in LAKES, or under MARINE conditions. Used in the manufacture of cement and in brickmaking.

Marsh. WETLAND whose dominant VEGETATION is grass. Marshes generally occur in the middle LATITUDES at the MOUTHS OF RIVERS, in estuaries and LAGOONS, and especially if there is a DELTA. Saltwater marshes

Typical marshland. (PhotoDisc)

are covered in a thick mat of sedges and similar plants and periodically are flooded by TIDES. Low ELEVATION and poor DRAINAGE provide marsh conditions where highly productive ECOSYSTEMS develop. Similar wetlands with tree vegetation are called SWAMPS. The Florida Everglades are a combination of marsh and swamp. Other small freshwater marsh areas are found in REGIONS previously covered by CONTINENTAL GLACIERS. On tropical COASTS, the wetlands are MANGROVE SWAMPS.

Mass balance. Summation of the net gain and loss of ice and SNOW mass on a GLACIER in a year.

Mass extinction. Die-off of a large percentage of species in a short time.

Mass wasting. Downslope movement of Earth materials under the direct influence of GRAVITY.

Massif. French term used in geology to describe very large, usually IGNEOUS INTRUSIVE bodies.

Material culture. Visible and tangible products or objects made and used by a particular group; includes clothing, weapons, household items, tools, and buildings.

Mean sea level. Average height of the SEA surface over a multiyear time span, taking into account STORMS, TIDES, and SEASONS.

Meander. U-shaped bend in a RIVER. MEANDERING RIVERS generally flow across a FLOODPLAIN that has been built up of ALLUVIUM deposited by the STREAM. An extremely tight meander is called a gooseneck; it is likely to become a cutoff, or OXBOW LAKE, after a FLOOD. Tectonic UP-LIFT can cause a river to continue downcutting along its meandering course, producing incised or entrenched meanders. The word comes from the winding, meandering river in Turkey that the Romans called Menderes.

Meandering river. RIVER confined essentially to a single CHANNEL that transports much of its SEDIMENT load as fine-grained material in SUS-PENSION.

A meandering river. (PhotoDisc)

Measurement, systems of. The imperial system of measurement used in the United States was brought by the British in the seventeenth century. Distances are measured in miles, feet, or inches; weights in tons, pounds, and ounces; volume in gallons, quarts, and pints. In most countries of the world, measurements are made in the International System of Units, or metric system. This system, which uses decimal fractions or units of ten, developed after the French Revolution. The unit of distance, the meter, was defined as one-ten-millionth of the length of the MERIDIAN passing through Paris. The unit of weight, the gram, was defined as the weight of one cubic centimeter of water at 39 DEGREES Fahrenheit (4 degrees Celsius). A liter was defined as the volume of a cube with a side of 10 centimeters. The standards were revised and expanded, starting in 1960, so that 1 meter is now defined, in the International System, as the distance traveled by light in a vacuum in 1/299,792,458 second.

Mechanical weathering. Another name for PHYSICAL WEATHERING, or the breaking down of ROCK into smaller pieces.

Mechanization. Replacement of human labor with machines. Mechanization occurred in AGRICULTURE as tractors, reapers, picking machinery, and similar technological inventions took the place of human farm labor. Mechanization in industry was part of the INDUSTRIAL REVOLUTION, as spinning and weaving machines were introduced into the textile industry.

Medical geography. Branch of geography specializing in the study of health and disease, with a particular emphasis on the areal spread or DIFFUSION

Example of mechanical weathering and rock uplifted by tree roots. (U.S. Geological Survey)

of disease. The spatial perspective of geography can lead to new medical insights. Geographers working with medical researchers in Africa have made great contributions to understanding the role of disease on that CONTINENT. John Snow's studies of the origin and spread of cholera in London in 1854 mark the beginnings of medical geography.

Mediterranean climate. Midlatitude CLIMATE characterized by wet winters and long dry summers. The climate predominates around the Mediterranean Sea, as well as in small parts of other CONTINENTS, including California, central Chile, the southernmost part of South Africa, and the southwest corner of Australia. These are all west COASTS of continents.

Megacity. Term for the world's largest URBAN AREAS, generally a conurbation with a POPULATION of more than five million. METROPOLIS is an alternative term.

Megalopolis. Conurbation formed when large cities coalesce physically into one huge built-up area. Originally coined by the French geographer Jean Gottman in the early 1960's for the northeastern part of the United States, from Boston to Washington, D.C.

Meltwater. Water derived from the melting of GLACIER ice.

Mental map. Each person's conception of the world. Persons organize space according to their mental maps. They know how to get to school or work, or a movie theater, for example, without having to look at a MAP. They may be able to cross a CITY using a mental map of major streets or highways, without actually knowing the details of the spaces traversed.

Mercalli scale. Qualitative SCALE used to describe EARTHQUAKE intensity before the creation of the RICHTER SCALE. The violence of SEISMIC shaking is given a number based on a description of the effects. The Mercalli scale uses Roman numerals I through XII for earthquakes ranging from "detected only by SEISMOGRAPHS" to "catastrophic."

Mercator projection. See CYLINDRICAL PROJECTION.

Meridian. Line of LONGITUDE.

Mesas in the Grand Canyon. (PhotoDisc)

Mesa. Flat-topped HILL with steep sides. EROSION removes the surrounding materials, while the mesa is protected by a cap of harder, more resistant ROCK. Usually found in arid REGIONS. A larger LANDFORM of this type is a PLATEAU; a smaller feature is a BUTTE. The Colorado Plateau and Grand Canyon in particular are rich in these landforms. From the Spanish word for table.

Mesosphere. Atmospheric layer above the STRATOSPHERE where TEMPERATURE drops rapidly.

Mesozoic era. Middle of the three ERAS that constitute the PHANEROZOIC EON (the last 544 million years), which encompasses three geologic PERIODS—the TRIASSIC, the JURASSIC, and the CRETACEOUS—and represents Earth history between about 250 and 65 million years ago.

Mestizo. Person of mixed European and Amerindian ancestry, especially in countries of LATIN AMERICA.

Metamorphic rock. Any ROCK whose mineralogy, MINERAL chemistry, or TEXTURE has been altered by heat, pressure, or changes in composition; metamorphic rocks may have IGNEOUS, SEDIMENTARY, or other, older metamorphic rocks as their precursors.

Metamorphic zone. Areas of ROCK affected by the same limited RANGE of TEMPERATURE and pressure conditions, commonly identified by the presence of a key individual MINERAL or group of minerals.

Metamorphism. Alteration of the mineralogy and TEXTURE of ROCKS because of changes in pressure and TEMPERATURE conditions or chemically active fluids.

Meteor. METEOROID that enters the ATMOSPHERE of a PLANET and is destroyed through frictional heating as it comes in contact with the various gases present in the atmosphere.

Meteor shower. Annual passage of Earth through a cometary wake or debris field, causing a METEOR display as COMET dust particles burn up in the upper ATMOSPHERE.

Meteoric water. Water that originally came from the ATMOSPHERE, perhaps in the form of rain or SNOW, as contrasted with water that has escaped from MAGMA.

Meteorite. Fragment of an ASTEROID that survives passage through the ATMOSPHERE and strikes the surface of the earth.

Meteoroid. Small planetary body that enters Earth's ATMOSPHERE because its path intersects the earth's ORBIT. Friction caused by the earth's atmosphere on the meteoroid creates a glowing METEOR, or "shooting star." This is a common phenomenon, and most meteors burn away completely. Those that are large enough to reach the ground are called METEORITES.

Meteorology. Study of short-term variations in the earth's ATMOSPHERE, particularly in the TROPOSPHERE. DAY-to-day changes in TEMPERATURE, HUMIDITY, PRECIPITATION, and pressure form the basis for meteorology. WEATHER forecasters use meteorological techniques. In contrast, CLIMATOLOGY is the description and analysis of CLIMATE, based on the study of long-term behavior of atmospheric variables.

Metropolis. Large CITY with its suburbs. From the Greek word for "mother city."

Metropolitan area. In general terms, a central CITY and the contiguous built-up area, together with the surrounding nonurban area that is economically tied to the central city. For statistical and CENSUS purposes, there exist formal definitions of a metropolitan area, which have been changed over time. In the year 2000, the U.S. Office of Management and Budget defined a metropolitan area (MA) as a CORE REGION containing a large POPULATION nucleus, together with adjacent communities having a high degree of economic and social integration with that core. MAs include metropolitan statistical areas (MSAs), consolidated metropolitan statistical areas (CMSAs), and primary metropolitan statistical areas (PMSAs). An MSA was defined as one city with 50,000 or more inhabitants, or a Census Bureau-defined urbanized area (of at least 50,000 inhabitants) and a total metropolitan population of at least 100,000 (75,000 in New England). An area that meets these requirements for recognition as an MSA and also has a population of one million or more may be recognized as a CMSA if separate component areas can be identified within the entire area by meeting statistical criteria specified in the standards, and local opinion indicates there is support for the component areas. If recognized, the component areas are designated PMSAs, and the entire area becomes a CMSA. PMSAs, like the CMSAs that contain them, are composed of entire counties, except in New England, where they are composed of cities and TOWNS. If no PMSAs are recognized, the entire area is designated as an MSA. In June, 1999, there were 258 MSAs, and 18 CMSAs comprising 73 PMSAs in the United States. In addition, there were 3 MSAs, 1 CMSA, and 3 PMSAs in Puerto Rico.

Microburst. Brief but intense downward WIND, lasting not more than fifteen minutes over an area of 0.6 to 0.9 square mile (1.5-8 sq. km.). Usually associated with THUNDERSTORMS, but are quite unpredictable. The sudden change in wind direction associated with a microburst can create wind shear that causes airplanes to crash, especially if it occurs during takeoff or landing. See also MACROBURST.

Microclimate. CLIMATE of a small area, at or within a few yards of the earth's surface. In this REGION, variations of TEMPERATURE, PRECIPITATION, and moisture can have a pronounced effect on the bioclimate, influencing the growth or well-being of plants and animals, including humans. DEW or FROST, RAIN SHADOW effects, wind-tunneling between tall buildings, and similar phenomena are studied by microclimatologists. Horticulturists know the variations in aspect that affect INSOLATION and temperature, so that certain plants grow best on south-facing walls, for example. The growing of grapes for wine production is a major industry where microclimatology is essential. The study of microclimatology was pioneered by the German meteorologist Rudolf Geiger.

Microcontinent. Independent LITHOSPHERIC PLATE that is smaller than a CONTINENT but possesses continental-type CRUST. Examples include Cuba and Japan.

Microstates. Tiny countries. In 2000, seventeen independent countries each had an area of less than 200 square miles (520 sq. km.). The

A true microstate, Vatican City is an independent country that occupies less than a fifth of a square mile within the Italian city of Rome. (PhotoDisc)

smallest microstate is Vatican City, with an area of 0.2 square miles (0.5 sq. km.). The tiny PRINCIPALITY of Monaco has an area of 1.0 square miles (1.95 sq. km.). Other European microstates include San Marino, Liechtenstein, and Andorra. Most of the world's microstates are island NATIONS, including Nauru, Tuvalu, Marshall Islands, Saint Kitts and Nevis, Seychelles, Maldives, Malta, Grenada, Saint Vincent and the Grenadines, Barbados, Antigua and Barbuda, and Palau.

Middle atmosphere. General term encompassing the STRATOSPHERE and the MESOSPHERE.

Mid-ocean ridge. Continuous mountain RANGE of underwater VOLCA-NOES located along the center of most OCEAN BASINS; volcanic ERUP-TIONS along these RIDGES drive SEAFLOOR SPREADING.

Migration. Change in PLACE of residence. Human migration used to be regarded as implying the intention of permanent residence at the destination, but this idea is breaking down as a result of modern transport. Migration is usually VOLUNTARY, and most twentieth century immigrants moved from low-income countries to high-income countries as economic immigrants. Migration can be FORCED, as in the case of political refugees fleeing repressive governments. These are examples of INTERNATIONAL MIGRATION. Demographers also study INTERNAL

MIGRATION, within a single COUNTRY, such as the SUN BELT migration in the United States.

Mineral. Substance that occurs naturally and has a unique chemical composition and a distinct crystal structure. Most minerals occur in compounds, but some metallic minerals occur as elements, such as copper and gold. ROCKS are composed of combinations of various minerals. The most common minerals on Earth are silicates.

Copper ore in its native state. (U.S. Geological Survey)

Mineral species. Mineralogic division in which all the varieties in any one species have the same basic physical and chemical properties.

Mineral variety. Division of a MINERAL SPECIES based upon color, type of optical phenomenon, or other distinguishing characteristics of appearance.

Miocene epoch. Geological EPOCH of the TERTIARY PERIOD in the CENOZOIC ERA, beginning about 26 million years ago.

Mist. Tiny water droplets—having a diameter of less than two hundred microns—held suspended in AIR. Visibility is impaired by the thin gray mist but remains above 0.6 mile (1 km.). When visibility is less than 0.6 mile, the condition is called FOG. A cold dense combination of fog and

Sunset mist in Florida. (Visit Florida)

rain DRIZZLE, often encountered in Scotland and similar cold CLIMATE REGIONS, is called scotch mist.

Model. Scientific term for a hypothetical description of an idea or phenomenon that explains its characteristics in a way that makes the model useful for further study of its characteristics. Examples in this glossary include CYCLE OF EROSION and DEMOGRAPHIC TRANSITION.

Monadnock. Isolated HILL far from a STREAM, composed of resistant BEDROCK. Monadnocks are found in humid temperate REGIONS. A similar LANDFORM in an arid region is an INSELBERG.

Monarchy. System or rule by a single person or sovereign ruler. The position is hereditary, as opposed to REPUBLICS, where the head of STATE is elected. Monarchs used to claim that they were appointed by God; this is referred to as the Divine Right of Kings. The Japanese held the concept of imperial divinity until 1945. As NATION-STATES evolved in Europe, some new monarchs were absolute rulers. In contrast, the power of the monarch in England was limited by the Parliament. Many modern monarchs have largely ceremonial roles.

Monogenetic. Pertaining to a volcanic ERUPTION in which a single vent is used only once.

Monsoon. Seasonal reversal of WIND. The largest monsoonal phenomenon is the Asian monsoon. In the summer wet SEASON, warm moist

AIR from over the Indian Ocean and South China Sea is drawn into the Asian CONTINENT, bring heavy rains, THUNDERSTORMS, and even TROPICAL CYCLONES. The wet monsoon ensures sufficient water for crops but can cause great loss of life through flooding and STORM SURGES. The dry or winter monsoon is marked by an outward flow of wind from Asia, bringing dry cooler conditions. Northern Africa and Northern Australia also experience monsoon conditions. From an Arabic word for season.

Moon. Any natural SATELLITE orbiting a PLANET. Earth has only one such satellite, which is called the Moon. Mercury and Venus have no moons, Mars has two small moons, Jupiter has sixteen known moons, Saturn has eighteen, Uranus has fifteen, Neptune has eight, and Pluto has one. Earth's Moon is 238,866 miles (384,400 km.) from Earth on average. The Moon is about one-third the size of the earth, with an equatorial diameter of 2,160 miles (3,476 km.). However, it is considerably lighter because its composition is less dense. The Moon revolves around the earth in an elliptical ORBIT every twenty-nine and a half DAYS. This corresponds to the time of one ROTATION of the Moon on its axis, so that only one side, or face, of the Moon can be seen from Earth. The Moon shines because of its ALBEDO, or reflected sunlight.

Moraine. Materials transported by a GLACIER, and often later deposited as a RIDGE of unsorted ROCKS and smaller material. Lateral moraine is found at the side of the glacier; medial moraine occurs when two gla-

Lateral and medial moraine in the French Alps. (Mark Twain, *A Tramp Abroad*, 1880)

ciers join. Other types of moraine include ABLATION moraine, ground moraine, and push, RECESSIONAL, and TERMINAL MORAINE.

Morphology. This word means structure or form. Geographers study the morphology of a COUNTRY, which explains many facts about its ECONOMY, CULTURE, historical geography, and politics. Generally, five morphologies are recognized: compact, elongated, fragmented, PERFORATED, and PROTRUDED (sometimes called prorupt).

Mountain. Tall LANDFORM, rising steeply above the surrounding COUNTRY, and with a comparatively narrow SUMMIT, or top. Most mountains occur in elongated groups, as mountain CHAINS or mountain RANGES. Mountains are produced by volcanic activity or by FOLDING or faulting of the earth's CRUST. A geologic term for a period of mountain-building is an OROGENY. In some countries, "mountains" are rigidly defined by their altitudes. Great Britain, for example, historically required landforms to be 1,000 feet (305 meters) high to be classified as mountains and classified lower-ELEVATION features as HILLS.

Mountains are high, massive landforms that rise steeply above the surrounding country and have comparatively narrow summits, or tops. Most mountains occur in elongated groups, as mountain chains or mountain ranges. (PhotoDisc)

Mountain belts. Products of PLATE TECTONICS, produced by the CONVERGENCE of crustal PLATES. Topographic MOUNTAINS are only the surficial expression of processes that profoundly deform and modify the CRUST. Long after the mountains themselves have been worn away, their former existence is recognizable from the structures that mountain building forms within the ROCKS of the crust.

Mountain glacier. GLACIER in a sloping VALLEY.

Mountain material. High-standing blocks of rugged RELIEF.

Mountain pass. See PASS.

Mouth of river. The PLACE where a RIVER enters a large body of standing water, such as the OCEAN or a LAKE. There much of the river's suspended load is deposited, often forming a DELTA. Many rivers enter the SEA in an ESTUARY, a long, narrow INLET where river water mixes with tidal waters.

Mudflow. General term for a flowing mass of predominantly fine-grained earth material that possesses a high degree of fluidity during movement.

Mulatto. Person of mixed African and European ancestry.

Multiculturalism. Government policy that enables and encourages ETHNIC GROUPS to retain their distinctive CULTURE and identity, instead of being assimilated into the larger dominant culture of the society.

Multinational corporation. Organization that engages in economic activities such as mining, AGRICULTURE, manufacturing, and marketing in more than one COUNTRY, affecting the economies of those countries.

Municipality. Any URBAN political unit, such as a TOWN or CITY.

Nappe. Huge sheet of ROCK that was the upper part of an overthrust fold, and which has broken and traveled far from its original position due to the tremendous forces. The Swiss Alps have nappes in many LOCATIONS.

Narrows. STRAIT joining two bodies of water.

Nation. Term originally meaning all the citizens of a REGION, sharing cultural traits such as a common LANGUAGE, RELIGION, and ethnicity. In the times of empires in Europe, the nation had no political meaning, since political allegiance was to the monarch or emperor, and religious allegiance was to the pope. One empire would include dozens of different cultural groups, or nations. The concept of NATIONALISM arose in the nineteenth century, when various nations wanted to occupy and control their own STATES, leading to the creation of modern NATION-STATES. The establishment of the League of Nations and the United Nations reflects the growth of nationalism in the twentieth century. The term nation now commonly is used to mean the state, or political entity. Modern countries are seldom nations in the older sense, since they rarely have a homogeneous POPULATION composition, owing to IMMIGRATION or BOUNDARY changes. The word still is used in the original sense to refer to groups such as the Navaho Nation.

Nation-state. Political entity comprising a COUNTRY whose people are a national group occupying the area. The concept originated in eighteenth century France; in practice, such cultural homogeneity is rare today, even in France.

National park. Designation given to land set aside by a national government for special protection. National parks tend to have unique qualities, such as spectacular scenery, unusual land formations, or endangered species of plants or animals in need of protection.

Nationalism. Feeling of belonging to a NATION, or a group of people with a common heritage and CULTURE. The rise of nationalism led to separatist movements and uprisings and the formation of new STATES in Europe throughout the nineteenth century, and in Africa in the twentieth century.

Native Americans. Widely accepted term for the native peoples of North America, especially those of the United States. Incorporates peoples also known as American ("Red") Indians and Inuit (Eskimos). Peoples of Canada are often known as Native Canadians. See also AMERINDIANS.

Natural bridge. Bridge over an abandoned or active watercourse; in KARST TOPOGRAPHY, it may be a short CAVE or a remnant of an old, long cave.

Utah's Arches National Park is named after its many natural red sandstone bridges, carved by millions of years of erosion. (Corbis)

Natural gas. Flammable vapor found in SEDIMENTARY ROCKS, commonly, but not always, associated with CRUDE OIL; it is also known simply as gas or methane.

Natural hazard. Natural event that causes loss of human life and property and environmental destruction. Natural hazards include FLOODS, HURRICANES and TORNADOES, EARTHQUAKES, volcanic ERUPTIONS, and TSUNAMI. Flooding is the natural hazard that causes the greatest loss of life.

Natural increase, rate of. DEMOGRAPHIC MEASURE of POPULATION growth: the difference between births and deaths per year, expressed as a percentage of the POPULATION. The rate of natural increase for the United States in 2000 was 0.6 percent. In countries where the population is decreasing, the DEATH RATE is greater than the BIRTH RATE.

Natural levee. Low RIDGE deposited on the flanks of a RIVER during a FLOOD stage.

Natural resource. See RESOURCE.

Natural selection. Main process of biological evolution; the production of the largest number of offspring by individuals with traits that are best adapted to their ENVIRONMENTS.

Nautical mile. Standard MEASUREMENT at SEA, equalling 6,076.12 feet (1.85 km.). The mile used for land measurements is called a statute mile and measures 5,280 feet (1.6 km.).

Navigation. Originally, the science and art of finding a safe, short path across water, requiring determination of distance, speed, course, and position. The positions of the SUN, MOON, and stars guided early navigators. Instruments developed to aid navigators include the magnetic COMPASS and sextant. CHARTS, MAPS, and guidebooks were also valuable aids to navigation. The invention of the MARINE CHRONOMETER enabled precision in LONGITUDE. Radio position-finding and GLOBAL POSITIONING SYSTEM satellites are modern aids to navigation. Scientists also study the secrets of animal navigation. Migratory birds can navigate thousands of miles, even from one HEMISPHERE to another in the course of a year.

Neap tide. TIDE with the minimum RANGE, or when the level of the high tide is at its lowest.

Near-polar orbit. Earth ORBIT that lies in a plane that passes close to both the north and south POLES.

Nekton. PELAGIC organisms that can swim freely, without having to rely on OCEAN CURRENTS or WINDS. Nekton includes shrimp; crabs; oysters; MARINE reptiles such as turtles, crocodiles, and snakes; and even sharks; porpoises; and whales.

Net migration. Net balance of a COUNTRY or REGION's IMMIGRATION and EMIGRATION.

Névé. French term for closely packed SNOW, deep in snowfields, from which the AIR has largely been expelled through compression due to the weight of overlying snow. Névé is an intermediate form between snow and glacial ice. It is like a series of clear bluish bands, representing the different years of snow accumulation. FIRN is the German term for this material.

Niche. In an ecological ENVIRONMENT, a position particularly suited for its inhabitant.

Nickpoint. See KNICKPOINT.

NIMBY. Acronym for "not in my back yard," in reference to movements opposing certain DEVELOPMENTS, especially in suburban areas. NIMBY-ism is generally a neighborhood action by residents who want to prevent unwanted LAND USES, protect open space, and maintain low-density housing, and block the nearby LOCATION of low-income housing or waste treatment plants, or facilities such as prisons or rehabilitation projects.

Niña, La. See LA NIÑA.

Niño, El. See EL NIÑO.

Nomadism. Lifestyle in which pastoral people move with grazing animals along a defined route, ensuring adequate pasturage and water for their flocks or herds. This lifestyle has decreased greatly as countries discourage INTERNATIONAL MIGRATION. A more restricted form of nomadism is TRANSHUMANCE.

Nomenclature. Names and terms used in a classification system.

Nonrenewable resource. RESOURCE that is exhausted after use. Includes FOSSIL FUELS, such as OIL and COAL, because the time for their formation is so long, although they are part of a biogeochemical CYCLE. See also RENEWABLE RESOURCES.

Normal fault. FAULT in which the ROCK block on top of an inclined fracture surface, also known as a FAULT PLANE, slides downward.

Normal fault in sandy shale in Tennessee's Chilhowee Mountains. (U.S. Geological Survey)

North geographic pole. Northernmost REGION of the earth, located at the northern point of the PLANET's AXIS of ROTATION.

North magnetic pole. Small, nonstationary area in the Arctic Circle toward which a COMPASS needle points from any LOCATION on the earth.

North/south divide. Term deriving from previous centuries, when European powers controlled large colonial empires. Many of the colonies were located in the SOUTHERN HEMISPHERE, including South Africa, Australia, New Zealand, and South America. Now, the term is sometimes used to refer economically to the contrast between high-income and low-income economies. Geographically, it makes little sense to use the term north/south divide in this way, because Australia and New Zealand are in the Southern Hemisphere, but are similar economically to wealthy NORTHERN HEMISPHERE countries.

Northern Hemisphere. The half of the earth above the EQUATOR.

Notch. Erosional feature found at the base of a SEA CLIFF as a result of undercutting by WAVE EROSION, bioabrasion from MARINE organisms, and dissolution of ROCK by GROUNDWATER seepage. Also known as a nip.

Nuclear energy. ENERGY produced from a naturally occurring isotope of uranium. In the process of nuclear FISSION, the unstable uranium isotope absorbs a neutron and splits to form tin and molybdenum. This releases more neurons, so a chain reaction proceeds, releasing vast amounts of heat energy. Nuclear energy was seen in the 1950's as the energy of the future, but safety fears and the problem of disposal of radioactive nuclear waste have led to public condemnation of nuclear power plants. Nevertheless, France generates more than half its power from nuclear energy. The alternative method of nuclear energy pro-

Nuclear power plant cooling tower. (PhotoDisc)

duction is nuclear FUSION, the energy released when two smaller atomic nuclei fuse into one larger nucleus. Isotopes of hydrogen are the fuel. Fusion occurs naturally in the SUN, but humans have not harnessed nuclear fusion, except to create the hydrogen bomb.

Nuée ardente. Hot cloud of ROCK fragments, ASH, and gases that suddenly and explosively erupt from some VOLCANOES and flow rapidly down their slopes.

Numerical weather prediction. System whereby mathematical equations are used, with the aid of computers, to describe and predict atmospheric processes.

Nunatak. Isolated MOUNTAIN PEAK or RIDGE that projects through a continental ICE SHEET. Found in Greenland and Antarctica.

Nunatak surrounded by a moraine in the Alaska Gulf region. Jefferies Glacier is to the left. (U.S. Geological Survey)

Oasis. Area surrounded by DESERT, where a permanent water supply, usually from an AQUIFER, permits AGRICULTURE.

Obduction. Tectonic collisional process, opposite in effect to SUBDUCTION, in which heavier OCEANIC CRUST is thrust up over lighter CONTINENTAL CRUST.

Oblate sphere. Flattened shape of the earth that is the result of ROTATION.

Occidental. Word meaning western. It is the opposite of oriental.

Occultation. ECLIPSE of any astronomical object other than the SUN or the MOON caused by the Moon or any PLANET, SATELLITE, or ASTEROID.

Ocean. Large body of water contained in an OCEAN BASIN. Oceans cover just over 70 percent of the earth's surface. The presence of so much liquid water serves to maintain moderate TEMPERATURES. The largest ocean, the Pacific, covers almost one-third of the earth. Other oceans are the Atlantic, Indian, Arctic, and Antarctic or Southern Ocean. The edge of an ocean basin is marked by the CONTINENTAL SHELF. A small, partially enclosed area of an ocean is a SEA.

Ocean basins. Large worldwide depressions that form the ultimate RESERVOIR for the earth's water supply.

Ocean circulation. Worldwide movement of water in the SEA.

Ocean current. Predictable circulation of water in the OCEAN, caused by a combination of WIND friction, Earth's ROTATION, and differences in TEMPERATURE and density of the waters. The five great oceanic circulations, known as GYRES, are in the North Pacific, North Atlantic, South Pacific, South Atlantic, and Indian Oceans. Because of the CORIOLIS EFFECT, the direction of circulation is CLOCKWISE in the NORTHERN HEMISPHERE and COUNTERCLOCKWISE in the SOUTHERN HEMISPHERE, except in the Indian Ocean, where the direction changes annually with the pattern of winds associated with the Asian MONSOON. Currents flowing toward the EQUATOR are cold currents; those flowing away from the equator are warm currents. An important current is the warm Gulf Stream, which flows north from the Gulf of Mexico along the East Coast of the United States; it crosses the North Atlantic, where it is called the North Atlantic Drift, and brings warmer conditions to the western parts of Europe. The West Coast of the United States is affected by the cool, south-flowing California Current. The cool Humboldt, or Peru, Current, which flows north along the South American coast, is an important indicator of whether there will be an EL NIÑO event. Deep currents, below 300 feet (100 meters), are extremely complicated and difficult to study.

Oceanic crust. Portion of the earth's CRUST under its OCEAN BASINS.

Oceanic island. ISLANDS arising from seafloor volcanic ERUPTIONS, rather than from continental shelves. The Hawaiian Islands are the best-known examples of oceanic islands.

Oceanography. The science of Earth's oceans and SEAS. Physical oceanography deals with the study of seawater—TEMPERATURE, density, WAVES, and CURRENTS. Chemical oceanography deals with the chemistry of biogeochemical CYCLES of the OCEANS. MARINE geology is the study of the features of the OCEAN BASINS. Biological oceanography is the study of marine ECOLOGY, or plants and animals of the oceans.

Off-planet. Pertaining to REGIONS off the earth in orbital or planetary space.

Offshore financial centers. The global financial network is concentrated in METROPOLITAN centers, but some investors and institutions place a high value on secrecy of accounts or on sheltering RESOURCES from taxation. To meet these needs, some small countries or MICROSTATES have developed as offshore financial centers. These include the Bahamas, Vanuatu, Cayman Islands, and Bahrain, as well as the mainland countries of Luxembourg, Liechtenstein, and Belize.

Oil. Greasy substance that remains liquid at room TEMPERATURE and is insoluble in water. Oils can be obtained from plants and seeds or from the bodies of animals, but the most economically important oil today is MINERAL oil or PETROLEUM, sometimes called CRUDE OIL. This is a product created millions of years ago from the bodies of MARINE organisms that were incorporated into layers of SEDIMENTARY ROCKS. The petroleum migrated through PERMEABLE rocks to form series of RESERVOIRS that constitute an oil field. Oil is the most important of the FOSSIL FUELS.

Oligocene epoch. Geological PERIOD about 38 million years ago in the TERTIARY PERIOD of the CENOZOIC ERA.

Oort Cloud. Reservoir of long-period COMETS that exist in a spherical DISTRIBUTION far beyond the outer planetary ORBIT of the SOLAR SYSTEM. The study of this vast REGION gives scientists a better understanding of the origin of the SUN and the PLANETS.

Orbit. The path followed by an astronomical body as it moves around an attracting body. In our SOLAR SYSTEM, PLANETS move in orbits around the SUN. Smaller bodies, or SATELLITES, move in orbits around the planets. The MOON orbits the earth. The shape of an orbit is elliptical, giving rise to the earth's PERIHELION and APHELION.

Order. Group of closely related genera; in mammals, orders include the rodents, bats, and whales.

Ordovician epoch. Time PERIOD covering the interval from 505 to 438 million years ago; follows the CAMBRIAN, which covers the interval from 570 to 505 million years ago.

Ore. Type of ROCK containing MINERALS in such a concentration that mining is economically feasible. Hematite is a common ore from which iron is extracted. Galena is the principal ore for zinc. The ore is always mixed with large amounts of worthless materials known as GANGUE, so that separation is necessary to recover the mineral.

Ore deposit. Natural accumulation of MINERAL matter from which the owner expects to extract a metal at a profit.

Orient. Old European term meaning "east," for Asia.

Orogenesis. Process of mountain-RANGE formation.

Orogenic belt. MOUNTAIN BELT composed of a core of METAMORPHIC and PLUtonic ROCKS and an adjacent THRUST BELT.

Orogeny. MOUNTAIN-building episode, or event, that extends over a period usually measured in tens of millions of years; also termed a revolution.

Orographic precipitation. Phenomenon caused when an AIR mass meets a topographic barrier, such as a mountain RANGE, and is forced to rise; the air cools to saturation, and orographic precipitation falls on the WINDWARD side as rain or SNOW. The lee side is a RAIN SHADOW. This effect is noticeable on the West Coast of the United States, which has

RAIN FOREST on the windward side of the MOUNTAINS and DESERTS on the lee.

Orography. Study of MOUNTAINS that incorporates assessment of how they influence and are affected by WEATHER and other variables.

Oscillatory flow. Flow of fluid with a regular back-and-forth pattern of motion.

Outback. Name by which Australians refer to any PLACE away from their cities. More specifically, the semiarid REGION west of the Great Dividing Range (Eastern Highlands), which covers about 80 percent of the CONTINENT, including most of Queensland, all of the Northern Territory, and all of Western Australia except the southwest corner.

Outer core. Zone in the body of the earth, located at depths of approximately 1,600 to 3,200 miles (2,900-5,100 km.), that is in a liquid state and consists of iron sulfides and iron oxides.

Overland flow. Flow of water over the land surface caused by direct PRECIPITATION.

Overurbanization. Growth of cities at such a rapid rate that they cannot sustain job creation and housing construction. Rapid in-migration from RURAL areas in low-income countries is the major cause of overurbanization. High homelessness and high unemployment result from overurbanization. URBAN POPULATION growth of this kind leads to many slums, shanties, and SQUATTER SETTLEMENTS that are illegally occupied and have few services.

Oxbow lake. LAKE created when floodwaters make a new, shorter CHANNEL and abandon the loop of a MEANDER. Over time, water in the oxbow lake evaporates, leaving a dry, curving, low-lying area known as a meander scar. Oxbow lakes are common on FLOODPLAINS. Another name for this feature is a cut-off.

Oxidation. Common chemical reaction in which elements are combined with oxygen—for example, the burning of PETROLEUM, wood, and COAL; the rusting of metallic iron; and the metabolic RESPIRATION of organisms.

Ozone. Gas containing three atoms of oxygen; it is highly concentrated in a zone of the STRATOSPHERE.

Ozone hole. Decrease in the abundance of ANTARCTIC OZONE as sunlight returns to the POLE in early springtime

Ozone layer. Narrow band of the STRATOSPHERE situated near 18 miles (30 km.) above the earth's surface, where molecules of OZONE are concentrated. The average concentration is only one in four million, but this thin layer protects the earth by absorbing much of the ultraviolet light from the SUN and reradiating it as longer-wavelength radiation. Scientists were disturbed to discover that the ozonosphere was being destroyed by photochemical reaction with CHLOROFLUOROCARBONS (CFCs). The OZONE HOLES over the South and North Poles negatively affect several animal species, including humans; skin cancer risk is increasing rapidly as a consequence of depletion of the ozone layer. Stratospheric ozone should not be confused with ozone at lower levels, which is a result of PHOTOCHEMICAL SMOG. Also called the ozonosphere.

P wave. Fastest elastic wave generated by an EARTHQUAKE or artificial ENERGY source; basically an acoustic or shock wave that compresses and stretches solid material in its path.

Pacific Rim. Group of countries with COASTLINES on the Pacific Ocean. In seeking to strengthen economic ties, these countries emphasize TRADE across the Pacific, instead of older trade links with Western Europe. Countries include the United States, Canada, Japan, Korea, China, Philippines, Australia, New Zealand, and Chile. The organization APEC was formed to exploit the growing economic strength of some Asian economies in Pacific Rim countries.

Pacific Ring of Fire. See RING OF FIRE.

Paddies. Rice fields, especially in Asian countries. The fields are small; the land must be level, with an impermeable SUBSOIL. Paddies are enclosed by low earth walls, so that they can retain the water required for flooding throughout the growing season. This type of rice growing is wet-rice cultivation. Many hillsides in Asian countries have been laboriously terraced to create small paddies on steep slopes.

Pahoehoe. See ROPY LAVA.

Paleobiogeography. Study of the geographic DISTRIBUTION of past life-forms.

Paleobiology. Study of the most ancient life-forms, typically through the examination of microscopic FOSSILS.

Paleoceanography. Study of the history of the OCEANS of the earth, ancient SEDIMENT DEPOSITION patterns, and OCEAN CURRENT positions compared to ancient CLIMATES.

Paleodepth. Estimate of the water depth at which ancient seafloor SEDIMENTS were originally deposited.

Paleomagnetism. Study of MAGNETISM preserved in ROCKS, which provides evidence of the history of Earth's MAGNETIC FIELD and the movements of CONTINENTS.

Paleontology. Study of ancient life; invertebrate paleontologists study FOSSIL invertebrate animals, vertebrate paleontologists study fossil vertebrates, and micropaleontologists study microfossils.

Paleozoic era. ERA that began about 543 million years ago and ended 245 million years ago; it includes six PERIODS: the CAMBRIAN, the ORDOVICIAN, the Silurian, the Devonian, the CARBONIFEROUS, and the PERMIAN.

Pandemic. Epidemic that spreads through a large area, sometimes even of worldwide proportions, leading to the deaths of millions of humans. The Black Death, or plague, that affected Europe in the fourteenth century was a pandemic in which it is estimated that twenty-five million people died. The Spanish Influenza Epidemic, which spread around the world during 1918 and 1919, was a pandemic, causing more than thirty million deaths. The current spread of AIDS/HIV might be described as a pandemic.

Pangaea. Name used by Alfred Wegener for the SUPERCONTINENT that broke apart to create the present CONTINENTS.

Paradigm. Pattern of scientific research and investigation that prevails in any discipline over time. A new way of conducting research or a new philosophical approach leads to a paradigm shift.

Parallel. Line of LATITUDE. One of a series of imaginary lines that extend around the earth parallel to the EQUATOR. Parallels are numbered from zero to ninety DEGREES north or south. The forty-ninth parallel forms part of the BORDER between Canada and the United States.

Parasitic cone. Small volcanic cone that appears on the flank of a larger VOLCANO, or perhaps inside a CALDERA.

Parish. Administrative subdivision of a British COUNTY or a division of the U.S. STATE of Louisiana that corresponds to the counties of other states.

Particulate matter. Mixture of small particles that adversely affect human health. The particles may come from smoke and DUST and are in their highest concentrations in large URBAN AREAS, where they contribute to the "DUST DOME." Increased occurrences of illnesses such as asthma and bronchitis, especially in children, are related to high concentrations of particulate matter.

Pass. Lower section between MOUNTAINS that enables people to travel across the mountain RANGE. It may be a saddle or, more commonly, a GORGE. The most famous is the Khyber Pass, on the BORDER between Afghanistan and Pakistan, which has been the entryway for numerous invasions of the Indian SUBCONTINENT. The Simplon Pass allowed traverse of the Swiss Alps between northern and southern Europe, but it has been superseded by the Simplon Tunnel. The Donner Pass in the Sierra Nevada is named after the expedition leaders, George and Jacob Donner, whose party of immigrants was snowbound there in the winter of 1846-1847, leading some members to survive on the dead bodies of their companions. Also called a mountain pass.

Pastoralism. Type of AGRICULTURE involving the raising of grazing animals, such as cattle, goats, and sheep. Pastoral nomads migrate with their domesticated animals in order to ensure sufficient grass and water for the animals.

Paternoster lakes. Small circular LAKES joined by a STREAM. These lakes are the result of glacial EROSION. The name comes from the resemblance to rosary beads and the accompanying prayer (the Our Father).

Patriarchy. Society in which men dominate all aspects of life, which can mean the oppression and exploitation of women. Fathers dominate the household and family group, and adult men have absolute authority over the larger social group. The early studies of patriarchy used ancient Greece and Rome as models. In practice, however, women had considerable power in those societies. Ethnographers and anthropologists have shown that there are many types of social arrangements that are complex and cooperative, rather than dominated entirely by one sex. Most feminists are opposed to patriarchal arrangements.

Patterned ground. Networks of rocks brought to the surface in polygon patterns by freeze-thaw cycles in arctic environments. See also PERIGLACIAL; PERMAFROST.

Peak. MOUNTAIN, or part of a mountain, that has a sharply defined SUMMIT. Some mountains, such as Africa's Mount Kilimanjaro, have more than a single peak.

A peak is mountain, or part of a mountain, that has a sharply defined summit, in contrast to hills and mountains with smooth domes. (PhotoDisc)

Peat. Organic material formed in cool humid CLIMATES, where VEGETATION has accumulated and partially decomposed under boggy or waterlogged conditions. It is an early stage in the formation of COAL. Peat is found extensively on all northern CONTINENTS and can be used as a fuel if it is dried before burning. Peat has been an important source of ENERGY in Ireland.

Pebble. On the SCALE developed by C. K. Wentworth, a pebble is a particle or piece of ROCK with a diameter between .16 and 2.5 inches (4 and 64 millimeters). Coarser material is cobbles; finer material is GRANULES. Pebbles are an important product of FLUVIAL EROSION. They are part of the formation of POTHOLES.

Pedestal rock. Rock that has assumed the shape of a pedestal as a result of unique shaping processes caused by WIND.

Pediment. In DESERT REGIONS, there is often a gently sloping BEDROCK surface extending along the base of a mountain RANGE, sometimes covered with a thin layer of ALLUVIUM. This EROSION surface is a special case of a PIEDMONT, which was given the name pediment. An extensive pediment has been called a pediplain.

Pediplain. See PEDIMENT.

Pedology. Scientific study of SOILS.

Pelagic. Relating to life-forms that live on or in open SEAS, rather than waters close to land.

Pele. The Hawaiian goddess of fire, attributed with the creation of the IS-LANDS of Hawaii. When strong WINDS blow pieces of LAVA from a lava fountain, drawing them out into long thin threads, the delicate volcanic forms are called Pele's hair.

Peneplain. In the geomorphic CYCLE, or cycle of LANDFORM development, described by W. M. Davis, the final stage of EROSION led to the creation of an extensive land surface with low RELIEF. Davis named this a peneplain, meaning "almost a plain." It is now known that tectonic forces are so frequent that there would be insufficient time for such a cycle to complete all stages required to complete this landform.

Peninsula. Narrow strip of land extending from the mainland into the SEA or OCEAN. From the Latin word meaning "almost an ISLAND."

Percolation. Downward movement of part of the water that falls on the surface of the earth, through the upper layers of PERMEABLE SOIL and ROCKS under the influence of GRAVITY. Eventually, it accumulates in the zone of SATURATION as GROUNDWATER.

Perennial lake. LAKE that contains water year-round. See also INTERMITTENT LAKE.

Perennial stream. RIVER that has water flowing in it throughout the year. See also INTERMITTENT STREAM.

Perforated state. STATE whose territory completely surrounds another state. The classic example of a perforated state is South Africa, within which lies the COUNTRY of Lesotho. Technically, Italy is perforated by the MICROSTATES of San Marino and Vatican City.

Periglacial. Landforms and processes found over one fifth the surface of the earth, along the margins of past and present GLACIERS. See also PERMAFROST; PATTERNED GROUND

Perihelion. Point in Earth's REVOLUTION when it is closest to the SUN (usually on January 3). At perihelion, the distance between the earth and the Sun is 91,500,000 miles (147,255,000 km.). The opposite of APHELION.

Period. Unit of geologic time comprising part of an ERA and subdivided, in decreasing order, into EPOCHS, ages, and chrons.

Periodicity. The recurrence of related phenomena at regular intervals.

Periphery. Geographic study of a COUNTRY or REGION can identify a core area or focus of human activity, where there is a concentration of POPulation, wealth, production, and consumption. A country's CORE RE-GION is usually its CAPITAL CITY and nearby region. Other parts of the country are the periphery and have the disadvantage of lower levels of DEVELOPMENT and investment. The concept can be applied to the whole world: There is a small core of countries with high national incomes and advanced living standards (Japan, the United States, Canada, and countries in Western Europe), and a large peripheral area of the rest of the world, where underdevelopment, hunger, poverty, and lack of education prevail. Countries in the periphery have no control of the global economic system and rely on the export of raw materials and cash crops to the core, or on the exploitation of their workers in low-wage labor.

Permafrost. Permanently frozen SUBSOIL. The condition occurs in perennially cold areas such as the ARCTIC. No trees can grow because their roots cannot penetrate the permafrost. The upper portion of the frozen SOIL can thaw briefly in the summer, allowing many smaller plants to thrive in the long daylight. Permafrost occurs in about 25 percent of the earth's land surface, and the condition even hampers construction in REGIONS such as Siberia and ARCTIC Canada. See also PATTERNED GROUND; PERIGLACIAL.

Exposed permafrost in Labrador's Katherine River Valley of the Torngat Mountains. (Geological Survey of Canada)

Permeable. Materials that can be penetrated by liquids or gases, such as porous ROCKS, are called permeable.

Permian period. Most recent PERIOD of the PALEOZOIC ERA, lasting from approximately 280 to 225 million years ago.

Perturb. To change the path of an orbiting body by a gravitational force.

Petrified wood. Form of FOSSIL wood in which all the original tissue and structure of the tree has been replaced by SILICA or calcite. It is produced when waterlogged tree trunks are buried in SAND, or also if trees are covered by VOLCANIC ASH. The Petrified Forest of Arizona is a famous REGION of the United States preserved as a NATIONAL PARK. Here the tree tissue has been replaced by chalcedony, a form of QUARTZ.

Petrified tree trunk, photographed around 1890, on Specimen Ridge, Yellowstone National Park. (U.S. Geological Survey)

Petrochemical. Chemical substance obtained from NATURAL GAS or PETROLEUM.

Petrography. Description and systematic classification of ROCKS.

Petroleum. Commonly used alternative term for CRUDE OIL; technically, the word refers to the mixture of complex hydrocarbons that can exist as gas (NATURAL GAS), liquid (OIL), and solid (bitumen). In the twentieth century, oil and natural gas were the most important FOSSIL FUELS. Petroleum formed from the altered remains of single-celled planktonic organisms accumulated deep in SEDIMENTARY ROCKS such as SHALE and CLAY, before migrating to porous RESERVOIR ROCKS. Most of the more than fifty thousand oilfields in the world are small. Although the United States is a major oil producer, it is also the world's largest consumer, so it is heavily reliant on imported oil. From a Greek word for rock oil.

Phanerozoic eon. PERIOD of geologic time with an abundant FOSSIL RECORD, extending from about 544 million years ago to the present.

Photochemical smog. Mixture of gases produced by the interaction of sunlight on the gases emanating from automobile exhausts. The gases include OZONE, nitrogen dioxide, carbon monoxide, and peroxyacetyl nitrates. Many large cities suffer from poor AIR quality because of photochemical smog. Severe health problems arise from continued exposure to photochemical smog.

Photometry. Technique of measuring the brightness of astronomical objects, usually with a photoelectric cell.

Photosynthesis. Process by which green plants capture light ENERGY—generally INSOLATION—and convert it into chemical compounds known as sugars (starches, hydrocarbons) that provide energy for the plants to grow. Hydrogen from water and carbon from the ATMOSPHERE combine with oxygen in these energy-rich compounds. Water and CARBON DIOXIDE from the atmosphere are essential for photosynthesis. Oxygen is the other product of photosynthesis. Without photosynthesis, life on Earth would be impossible. Plants, which rely on photosynthesis for growth, form the basis of any FOOD CHAIN, so organisms, including humans, would not survive without it. The opposite of photosynthesis is RESPIRATION.

Phylogeny. Study of the evolutionary relationships among organisms.

Phylum. Major grouping of organisms, distinguished on the basis of basic body plan, grade of anatomical complexity, and pattern of growth or development.

Physical geography. The study of the natural world, including GEOMORPHOLOGY, CLIMATOLOGY, BIOGEOGRAPHY, SOILS, and aspects of MARINE studies and environmental science. Physical geographers are especially interested in the relationship between humans and the natural world.

Physical weathering. The breaking down or disintegrating of ROCK into smaller pieces. In cold CLIMATES, or in high-ALTITUDE REGIONS where TEMPERATURES fall below zero at night, FROST WEDGING is an important form of physical weathering, which causes rocks to shatter as water freezes into ice in JOINTS in rocks. EXFOLIATION or the peeling off of sheets of rock through pressure release is another form of physical weathering.

Physiography. The PHYSICAL GEOGRAPHY of a PLACE—the LANDFORMS, water features, CLIMATE, SOILS, and VEGETATION.

Physiologic density. Measure of agricultural productivity—the number of people of a COUNTRY who are fed per unit area of ARABLE, or agricultural, land. Several countries have a much higher physiological density than the United States, but they also have either higher POPULATIONS, less arable land, or both.

Piedmont. LANDFORM at the foot of a MOUNTAIN. In arid REGIONS, it is easy to observe the piedmont angle—the sharp change in angle of slope between the flat DESERT PLAIN and the adjacent mountains. Piedmont can be an erosional surface, cut into a BEDROCK surface called a PEDIMENT, or a depositional surface, formed as STREAMS emerge from the mountains and deposit their load as ALLUVIAL FANS. A series of coalescing alluvial fans form a sloping piedmont called a *bajada.* The name "piedmont" comes from the Piedmont of northern Italy, which comprises the Po River Valley. In the United States, the long region of dissected PLATEAU lying just east of the Appalachian Mountains, from New Jersey to Alabama, is called the Piedmont; it contains some of the richest farmland of that COUNTRY.

Piedmont glacier. GLACIER formed when several ALPINE GLACIERS join together into a spreading glacier at the base of a MOUNTAIN or RANGE. The Malaspina glacier in Alaska is a good example of a piedmont glacier.

Alaska's Malaspina Glacier, with the Mount St. Elias range in the background. (U.S. Geological Survey)

Piedmont lake. LAKE formed when glacial MORAINE, or deposited material, DAMS up a STREAM flowing in a former glacial TROUGH. Several lakes forming the Lake District of England are piedmont lakes.

Pilgrimage. Journey to a sacred SITE or to a PLACE of religious importance, such as a shrine, undertaken by believers of that faith. The *hajj*, or pilgrimage to Mecca, is especially important for the RELIGION of Islam. Christian pilgrimages to Jerusalem have been made for almost two millennia. The most important site for Buddhist pilgrims is Bodh Gaya in India, where Prince Siddhartha attained enlightenment.

Pillow lava rocks from off the shore of the island of Hawaii. (National Oceanic and Atmospheric Administration)

Pillow lava. Substance formed when a VOLCANO emits fluid ROPY LAVA, also known as pahoehoe, into the SEA, where the rapid cooling forms a skin, producing small rounded shapes like pillows, one after another, in a budding process.

Place. In geographic terms, space that is endowed with physical and human meaning. Geographers study the relationship between people, places, and ENVIRONMENTS. The five themes that geographers use to examine the world are LOCATION, place, human/environment interaction, movement, and REGIONS.

Placer. Accumulation of valuable MINERALS formed when grains of the minerals are physically deposited along with other, nonvaluable mineral grains.

Plain. Area of land that has low RELIEF or is almost flat. Difference in ELEVATION on a plain is less than 325 feet (100 meters) and the slope angle is less than five DEGREES. Most plains were formed by DEPOSITION, especially by RIVERS during FLOOD. FLOODPLAINS became favored LOCATIONS early in human history because the flat land was suitable for AGRICULTURE, building construction, and ease of transport. Almost one-third of the earth's land surface is plains. The Great Plains of North America extend from the Gulf of Mexico to Hudson Bay. Plains cover Europe from Poland to the Ural Mountains in Russia. South America has extensive plains, especially in Brazil and Argentina.

Plane of the ecliptic. See ECLIPTIC, PLANE OF.

Planet. Celestial body that revolves in an ORBIT around a star. This definition excludes MOONS, METEOROIDS, and COMETS. Our SOLAR SYSTEM has nine planets, of which Earth is the third-closest to the SUN. The other planets (in order from the Sun) are Mercury, Venus, Mars, Jupiter, Saturn, Uranus, Neptune, and Pluto. The inner four planets are called TERRESTRIAL PLANETS because they are made of solid ROCKS; the next four are the Jovian, or giant, planets made of gas.

Planetary wind system. Global atmospheric circulation pattern, as in the BELT of prevailing westerly WINDS.

Plankton. Organisms living in FRESH WATER or OCEANS that are too tiny to swim against CURRENTS or WIND and so can only float. Plankton can be subdivided based on life-forms into phytoplankton, or plantlike plankton, and zooplankton, or animal-like plankton. The crustacean zooplankton known as krill are the most important part of the MARINE plankton, because they are the food source for the great whales. Plankton also can be classified based on size. Macroplankton are greater than 0.4 inch (1 centimeter) in length, microplankton can be as small as a twentieth of that size, and nannoplankton are ten times smaller still. Organisms that can swim are called NEKTON.

Plankton are marine organisms so small they cannot swim against currents or wind and must therefore merely float where the water takes them. Under certain conditions, the numbers of organisms can explode, creating what is called a plankton bloom that is visible from space, as is this example off the coast of Namibia in the South Atlantic Ocean in April, 1985. (Corbis)

Plant communities. See Biome.

Plantation. Form of AGRICULTURE in which a large area of agricultural land is devoted to the production of a single cash crop, for export. Many plantation crops are tropical, such as bananas, sugarcane, and rubber. Coffee and tea plantations require cooler CLIMATES. Formerly, slave labor was used on most plantations, and the owners were usually Europeans.

Plat. General term for any small piece of land. Also used for a MAP showing features within a MUNICIPALITY.

Plate. Relatively thin slab of crustal ROCK, either continental or oceanic, that moves over the face of the globe, driven by currents of circulating molten rock in the underlying MANTLE. See also TECTONIC PLATE.

Plate boundary. REGION in which the earth's crustal PLATES meet, as a converging (SUBDUCTION ZONE), diverging (MID-OCEAN RIDGE), TRANSFORM FAULT, or collisional interaction.

Plate tectonics. Theory proposed by German scientist Alfred Wegener in 1910. Based on extensive study of ancient geology, STRATIGRAPHY, and CLIMATE, Wegener concluded that the CONTINENTS were formerly one single enormous LANDMASS, which he named PANGAEA. Over the past 250 million years, Pangaea broke apart, first into LAURASIA and GONDWANALAND, and subsequently into the present continents. Earth scientists now believe that the earth's CRUST is composed of a series of thin, rigid PLATES that are in motion, sometimes diverging, sometimes colliding.

Plateau. Large area of flat land at high ELEVATION and surrounded by ESCARPMENTS. Plateaus and enclosed BASINS cover about 45 percent of the earth's land surface. The flat top often results from a hard layer of ROCK at the surface. The Colorado River has eroded the Grand Canyon into the Kaibab Plateau. A similar but smaller feature is a TABLELAND. The highest plateau is in Tibet (Xizang), where the elevation is around 14,500 feet (4,500 meters). A plateau can be formed by thermal expansion or UPLIFT by underlying hot MAGMA. The large plateaus of East Africa were formed in this way. Plateaus can also be formed by volcanic ERUPTION of huge amounts of LAVA, known as flood BASALT. The Deccan Plateau of India and the Columbian Plateau of the United States are good example of this type of plateau. Some plateaus are part of a mountain RANGE, for example, the ALTIPLANO in South America's Andes.

Playa. Shallow but broad SALINE LAKE, perhaps only a few centimeters deep, found in DRAINAGE BASINS in arid and semiarid REGIONS. A playa may dry up to form a salina, or SALT flat. Playas in the Western United States are sometimes called ALKALI FLATS.

Pleistocene. EPOCH of geological time extending from about 1.6 million years ago to about 10,000 years ago. Broadly speaking, the Pleistocene corresponded to the great ICE AGES of the earth, although it is now thought that the cooling began earlier. During the Pleistocene, almost one-third of the earth's land surface was covered by glacial ice. Now only about 10 percent is covered, mainly in Antarctica and Greenland. After the Pleistocene comes the HOLOCENE Epoch; preceding the Pleistocene was the PLIOCENE.

Plinian eruption. Rapid ejection of large volumes of VOLCANIC ASH that is often accompanied by the collapse of the upper part of the VOLCANO. Named either for Pliny the Elder, a Roman naturalist who died while observing the ERUPTION of Mount Vesuvius in 79 C.E., or for Pliny the Younger, his nephew, who chronicled the eruption.

Pliocene epoch. Geological EPOCH in TERTIARY PERIOD of the CENOZOIC ERA that began about 12 million years ago.

Plucking. Term used to describe the way glacial ice can erode large pieces of ROCK as it makes its way downslope. The ice penetrates JOINTS, other openings on the floor, or perhaps the side wall, and freezes around the block of stone, tearing it away and carrying it along, as part of the glacial MORAINE. The rocks contribute greatly to glacial ABRASION, causing deep grooves or STRIATIONS in some places. The jagged torn surface left behind is subject to further plucking. ALPINE GLACIERS can erode steep VALLEYS called glacial TROUGHS.

Plume. Expanded and cooled material pushed upward in the form of a fireball from the force of an ERUPTION or impact.

The plume of an erupting volcano is similar in appearance to that of an atomic bomb explosion. (PhotoDisc)

Plural society. Society in which more than one ETHNIC GROUP lives, with distinct separation of the different CULTURES. The United States and Canada are plural societies.

Pluton. Generic term for an IGNEOUS body that solidifies well below the earth's surface; PLUTONIC ROCKS are coarse-grained because they cool slowly.

Plutonic. IGNEOUS ROCKS made of MINERAL grains visible to the naked eye. These igneous rocks have cooled relatively slowly. GRANITE is a good example of a plutonic rock.

Pluvial period. Episode of time during which rains were abundant, especially during the last ICE AGE, from a few million to about ten thousand years ago.

Polar ice cap. Large sheet of ice, often more than a hundred square miles in size, that covers the polar portions of the Arctic Ocean and does not melt seasonally.

Polar stratospheric clouds. CLOUDS of ice crystals formed at extremely low TEMPERATURES in the polar STRATOSPHERE.

Polar vortex. Closed atmospheric circulation pattern around the South Pole that exists during the winter and early spring; atmospheric mixing between the polar vortex and REGIONS outside the vortex is slow. The low-pressure system has swirling WINDS at its boundaries.

Polarity. Orientation of the earth's MAGNETIC FIELD relative to the earth.

Polder. Lands reclaimed from the SEA by constructing DIKES to hold back the sea and then pumping out the water retained between the dikes and the land. Before AGRICULTURE is possible, the SOIL must be specially treated to remove the SALT. Some polders are used for recreational land; cities also have been built on polders. The largest polders are in the Netherlands, where the northern part, known as the Low Netherlands, covers almost half of the total area of this COUNTRY.

Pole. The ends of the earth's AXIS of ROTATION are termed the North and South Pole, respectively. The geographic pole does not correspond exactly with the MAGNETIC POLE, because the earth's MAGNETIC FIELD is in constant change. Norwegian explorer Roald Amundsen was the first to reach the South Pole, in 1911. American explorer Robert E. Peary claimed to have reached the North Pole in 1909. Richard E. Byrd flew over the North Pole in 1926.

Political geography. Study of spatial aspects of political processes, mainly at the international scale. It includes the spatial analysis of various political ideologies, BOUNDARY changes, forms of government, selection of CAPITAL cities, and relations between STATES.

Pollution. Environmental pollution is the introduction of unwanted and usually unhealthful materials into the ENVIRONMENT—the ATMOSPHERE, SOIL, or water. Some pollutants occur naturally: VOLCANOES and naturally occurring FOREST fires emit DUST and vapors into the atmosphere. Of greater concern is anthropogenic pollution, or human pollution of the environment. When the human POPULATION was small, this was not a problem, but six billion humans now produce vast amounts of human waste (sewage) and garbage. AIR POLLUTION is especially bad in many cities, such as Mexico City and Bangkok, where noxious gases are produced by vehicles and ENERGY sources. Human activities such as farming can increase dust particles, and human-caused forest fires are a major source of air pollution. Industrial products such as pesticides or CFCs (CHLOROFLUOROCARBONS) spread far from their source areas. Water pollution comes from sewage disposal and RUNOFF from cities, industries, and agricultural lands; THERMAL POLLUTION occurs from nuclear power plants. Land pollution generally involves solid waste, such as nonbiodegradable trash, but also includes chemicals such as DDT or PCBs.

Polygenetic. Pertaining to volcanism from several physically distinct vents or repeated ERUPTIONS from a single vent punctuated by long periods of quiescence.

Polygonal ground. Distinctive geological formation caused by the repetitive freezing and thawing of PERMAFROST.

Pool. Small but comparatively deep body of water, or a slow-moving part of a STREAM.

Popular culture. Term for the products and LANDSCAPES created, used, and enjoyed by people in their everyday lives. Popular culture is similar to folk culture and is the opposite of high culture.

Population. Human population refers to the number of people inhabiting an area, such as a COUNTRY. The population of the United States in May, 2000, was estimated to be 274,863,982. The population of the earth then was estimated at 6,072,255,639. In biology, population refers to the number of individuals of one species in a given area, for example the wolf population of Yellowstone National Park. Population size is limited by availability of food and water, disease, and other factors.

Population density. DEMOGRAPHIC MEASURE calculated as the total POPULATION of a COUNTRY or other territory divided by it total area. The resultant figure gives the number of residents per square mile or square kilometer. The United States had a population density of about 75 persons per square mile in the year 2000. Population density figures can be misleading, especially for large countries, since people are not distributed evenly throughout any country. In the United States, there are large clusters of dense population, such as the Los Angeles conurbation, where the population density is much higher then the average for the country; there also are huge REGIONS with almost no residents, such as in the STATES of Utah, Nevada, and Montana, where the population density is very low.

Population explosion. Great increase in the number of people in a short period, due to a large number of births with a high survival rate. The first human population explosion occurred about three thousand years ago, as a result of the AGRICULTURAL REVOLUTION. When many humans changed their lifestyle from HUNTING AND GATHERING to raising crops and domesticated animals, the increased food supply made this population explosion possible. The next great population explosion occurred in the eighteenth century, as a result of the INDUSTRIAL REVOLUTION in Europe. This explosion has continued. In the year 2000, there were more than six billion people on Earth, and this number was predicted to double in fifty years.

Population pyramid. Type of bar graph that displays the age and sex structure of a given POPULATION. A pyramid shape means that the sexes are evenly distributed; the greater numbers of the population are children, with correspondingly smaller numbers for each older age group; and there is continued population growth. Some rapidly growing countries have oversteepened population pyramids. Pyramids for countries with slow or no population growth have a smaller base than the middle sections that represent the adult population.

Porosity. The space in a ROCK or SOIL that is filled with AIR. This pore space occurs between the grains or crystals that make up the rock. Porosity is related to grain size, smoothness, and compaction of materials.

Port. PLACE on a COAST where ships can be securely anchored or tied up, safe from STORMS, while they load or unload cargo or passengers. As ships have become larger, many older ports have lost their economic function and large artificial HARBORS have been constructed in their place. The historic English Cinque Ports were Dover, Hastings, Hythe, New Romney, and Sandwich; only Dover remained important at the end of the twentieth century.

Possibilism. Concept that arose among French geographers who rejected the concept of ENVIRONMENTAL DETERMINISM, instead asserting that the relationship between human beings and the ENVIRONMENT is interactive.

Postindustrial economy. Concept introduced by American sociologist Daniel Bell, referring to the fact that the majority of the workforce in highly developed, high-income countries is employed in service industries. This is the result of increased disposable income and increased leisure time. Rapid growth of knowledge-based industries and information TECHNOLOGY has increased the importance of TERTIARY INDUSTRY and the QUATERNARY SECTOR, while the proportion of the workforce in SECONDARY INDUSTRY continues to decline. In such societies, education, health services, and the welfare state assume new dominance, while science and knowledge-based enterprises flourish.

Potable water. FRESH WATER that is being used for domestic consumption.

Potholes. Circular depressions formed in the bed of a RIVER when the STREAM flows over BEDROCK. The scouring of PEBBLES as a result of water TURBULENCE wears away the sides of the depression, deepening it vertically and producing a smooth, rounded pothole. (In modern parlance, the term is also applied to holes in public roads.)

Prairie. Flat PLAINS covered with grasses, found in North America. The annual RAINFALL decreases from east to west, and the VEGETATION changes correspondingly from tall-grass prairie through mixed prairie to short-grass prairie. The tall-grass prairie is an area of extremely rich SOILS, and the original vegetation has largely been cleared for grain farming of wheat or corn, or for URBAN purposes; the short-grass prairie is largely used for pastoral AGRICULTURE, especially raising cattle for beef. Prairie animals include the bison, wolf, prairie chicken, prairie dog, coyote, jackrabbit, and many birds. The Canadian PROVINCES of Alberta, Manitoba, and Saskatchewan are called the Prairie Provinces. The prairie BIOME is part of the middle LATITUDE GRASSLAND biome, which includes the Pampas of Uruguay and the grasslands of the Ukraine.

Prairie is the term applied to flat plains covered with grasses—a terrain that predominates throughout North America's Great Plains. (PhotoDisc)

Precambrian period. The oldest and longest time in Earth's geologic history. It began around 3.9 billion years ago—the age of the oldest known ROCKS on Earth—and continued to the beginning of the PALEOZOIC ERA, 543 million years ago. The name was given after the earliest FOSSILS were found in rocks in Wales (Cambria). It was assumed that life on Earth began in the CAMBRIAN PERIOD. Study in the twentieth century revealed that life has existed on Earth for perhaps 3.5 billion years. Because of greater knowledge, the Precambrian is now divided into the Archean and PROTEROZOIC EONS. The Precambrian period accounts for almost 80 percent of Earth's geologic history.

Precipitation. All water that falls from CLOUDS to the ground, whether in liquid or solid form. Water in the ATMOSPHERE collects around particles called CONDENSATION NUCLEI, forming cloud droplets that grow in size through collision and coalescence. The precipitation particles eventually become so large and heavy that they fall to the ground. Types of precipitation include SNOW, rain, SLEET, and hail. Snow is a solid, crystalline form of water. Rain is liquid water drops with diameters greater than 0.02 inch (0.5 millimeter). When the drops are smaller, the precipitation is usually called DRIZZLE. Sleet is frozen raindrops or partially melted snowflakes. Hail is balls or pieces of ice with a diameter of larger than 0.2 inch (5 millimeters). Small hail is sometimes called ice pellets.

Primary economic activity. Economic activities that derive their materials directly from the ENVIRONMENT. These include hunting, fishing, forestry, farming, mining, and quarrying. In low-income economies, the majority of the POPULATION is engaged in primary industry. In the poorest countries, this is usually SUBSISTENCE AGRICULTURE. Also called primary industries.

Primary minerals. MINERALS formed when MAGMA crystallizes.

Primary wave. Compressional type of EARTHQUAKE wave, which can travel in any medium and is the fastest wave.

Primate city. CITY that is at least twice as large as the next-largest city in that COUNTRY. The "law of the primate city" was developed by American geographer Mark Jefferson, to analyze the phenomenon of countries where one huge city dominates the political, economic, and cultural life of that country. The concept is easily understood when one thinks of Paris, a classic example of a primate city; London is another great primate city. The size and dominance of a primate city is a PULL FACTOR and ensures its continuing dominance. Not all countries have a primate city. The United States does not, because there are similar-sized agglomerations on the East Coast (New York) and the West Coast (Los Angeles), neither of which is the national CAPITAL city. It is not necessary for a primate city to be a national capital, but in practice this is the case. Australia is another country with no primate city. Instead, it has two large cities, Sydney and Melbourne, neither of which is the capital city.

Prime meridian. Line of LONGITUDE used as a reference for the geographic GRID. It is numbered zero and separates the EASTERN and WESTERN HEMISPHERES. Other MERIDIANS are numbered from 1 to 180 DEGREES, east and west, of the prime meridian. The prime meridian line is also called the Greenwich meridian because it runs through the former Royal Observatory at Greenwich, near London, England, connecting the North and South Poles. Although other cities, such as Paris and Washington, D.C., vied to be the LOCATION of the prime meridian, the decision was made in 1884 at the International Meridian

Conference in Washington, D.C. Great Britain then had the world's largest empire and the largest navy.

Principal parallels. The most important lines of LATITUDE. PARALLELS are imaginary lines, parallel to the EQUATOR. The principal parallels are the equator at ZERO DEGREES, the tropic of CANCER at 23.5 degrees North, the tropic of CAPRICORN at 23.5 degrees south, the Arctic Circle at 66.5 degrees north, and the Antarctic Circle at 66.5 degrees south.

Principality. Literally, the territory governed by a prince; any monarch's REALM.

Prorupt. See PROTRUDED.

Protectorate. COUNTRY that is a political DEPENDENCY of another NATION; similar to a COLONY, but usually having a less restrictive relationship with its overseeing power.

Proterozoic eon. Interval between 2.5 billion and 544 million years ago. During this PERIOD in the GEOLOGIC RECORD, processes presently active on Earth first appeared, notably the first clear evidence for PLATE TECTONICS. ROCKS of the Proterozoic eon also document changes in conditions on Earth, particularly an apparent increase in atmospheric oxygen.

Protruded. The MORPHOLOGY of a COUNTRY can be described as protruded when the main body of the country has a long thin extension stretching away from it. Thailand is a protruded country. The STATE of Oklahoma can be described as protruded. An alternative term is prorupt.

Province. Term used in some countries for internal administrative subdivisions. Canada, for example, has ten provinces. When South Africa was reorganized in 1994, it changed from four to nine provinces.

Psychrometer. Device used to measure and calculate the RELATIVE HUMIDITY of AIR. The sling psychrometer consists of two THERMOMETERS, one of which has its bulb wrapped in moistened cloth. The psychrometer is swung in the air, and the difference in TEMPERATURE of the two bulbs is calculated. This is compared with the psychrometric table, which gives the relative humidity value corresponding to those conditions.

Pull factors. Forces that attract immigrants to a new COUNTRY or LOCATION as permanent settlers. They include economic opportunities, educational facilities, land ownership, gold rushes, CLIMATE conditions, democracy, and similar factors of attraction.

Pumice. Light, porous ROCK of IGNEOUS origin. It is formed when ejected LAVA cools rapidly without crystallization. In this respect, pumice is similar to obsidian, or volcanic glass. Heating obsidian can produce pumice. The expulsion of gases in the lava causes the rock to swell, or froth, as it cools. Visually, pumice resembles some SEA sponges. It is pale gray to whitish in color and is so light that it floats in water. Countries around the Mediterranean Sea mine pumice commercially. Pumice from Peru is widely used in the fashion industry to create the "stone-washed" look of clothing items.

Push factors. Forces that encourage people to migrate permanently from their HOMELANDS to settle in a new destination. They include war, persecution for religious or political reasons, hunger, and similar negative factors.

Pyroclasts. Materials that are ejected from a VOLCANO into the AIR. Pyroclastic materials return to Earth at greater or lesser distances, depending on their size and the height to which they are thrown by the explosion of the volcano. The largest pyroclasts are volcanic bombs. Smaller pieces are volcanic blocks and scoria. These generally fall back onto the volcano and roll down the sides. Even smaller pyroclasts are LAPILLI, cinders, and VOLCANIC ASH. The finest pyroclastic materials may be carried by WINDS for great distances, even completely around the earth, as was the case with DUST from the Krakatoa explosion in 1883 and the early 1990's explosions of Mount Pinatubo in the Philippines.

Pyroclastic flow deposit. (U.S. Geological Survey)

Qanat. Method used in arid REGIONS to bring GROUNDWATER from mountainous regions to lower and flatter agricultural land. A qanat is a long tunnel or series of tunnels, perhaps more than a mile long. The word *qanat* is Arabic, but the first qanats are thought to have been constructed in Farsi-speaking Persia more than two thousand years ago. Qanats are still used there, as well as in Afghanistan and Morocco.

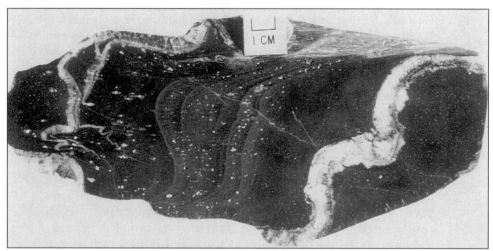

The simple oxide quartz (the lighter buckled veins in this rock sample) is the most common of the silicate minerals, which constitute 95 percent of Earth's crust. (Geological Survey of Canada)

Quartz. One of the most common MINERALS on the earth's surface; it occurs in many different forms, including agate, jasper, and chert.

Quaternary period. The shortest and youngest of the eleven PERIODS into which geologic time is divided. The Quaternary began around 1.6 million years ago and continues at the present. It comes after the TERTIARY PERIOD, which extended from 66.4 million years ago to 1.6 million years ago. The Quaternary is subdivided into the PLEISTOCENE and HOLOCENE EPOCHS. We are living in the Holocene Epoch.

Quaternary sector. Economic activity that involves the collection and processing of information. The rapid spread of computers and the Internet caused a major increase in the importance of employment in the quaternary sector. See also POSTINDUSTRIAL ECONOMY.

Radar imaging. Technique of transmitting radar toward an object and then receiving the reflected radiation so that time-of-flight MEASUREMENTS provide information about surface TOPOGRAPHY of the object under study.

Radial drainage. The pattern of STREAM courses often reveals the underlying geology or structure of a REGION. In a radial drainage pattern, streams radiate outward from a center, like spokes on a wheel, because they flow down the slopes of a VOLCANO.

Radiation. Transfer of ENERGY through a transparent medium, as occurs when the SUN warms the earth.

Radioactive minerals. MINERALS combining uranium, thorium, and radium with other elements. Useful for nuclear TECHNOLOGY, these minerals furnish the basic isotopes necessary not only for nuclear reactors but also for advanced medical treatments, metallurgical analysis, and chemicophysical research.

Radioactivity. ENERGY emitted spontaneously from certain types of ROCKS, through the decay of an unstable nucleus. The unstable materials that occur naturally on Earth include uranium-238, uranium-235, and thorium-232. Radioactivity has been harnessed to produce nuclear weapons and for nuclear power generation.

Radiocarbon dating. See CARBON DATING.

Radon gas. Radioactive gas and the heaviest of the noble gases. It is produced by the radioactive decay of radium, which is a natural decay product of the uranium found in various types of ROCKS. Trace amounts of radon seep from rocks and SOIL into the ATMOSPHERE and can become a health hazard in sufficient concentrations.

Rain forest. Dense evergreen FOREST with high annual RAINFALL. Most rain forests are in the TROPICS. A special rain forest is the temperate rain forest of the West Coast of the United States, where the redwoods and giant sequoias occur in small patches, now mostly protected. The tallest living trees in the world are located in this BIOME.

Most rain forests are in the Tropics; however, a special rain forest is the temperate redwood rain forest of the West Coast of the United States, where the redwoods and giant sequoias occur in small patches. (Digital Stock)

Rain forests are characterized by canopies of trees so dense that sunlight rarely reaches the ground. (PhotoDisc)

Rain forest, tropical. RAIN FOREST with monthly TEMPERATURES averaging greater than 65 DEGREES Fahrenheit (18 degrees Celsius) and monthly PRECIPITATION averaging more than 70 inches (1,800 millimeters) annually. The result is the most biologically diverse BIOME on Earth. The world's greatest expanse of tropical rain forest is in South America's Amazon Basin. Rain forests are also found in Africa, Central and South America, Asia, northern Australia, and Hawaii. Tropical rain forest covers about 7 percent of Earth's land surface, but it is being cleared at an alarming rate, especially in Brazil. This DEFORESTATION has been monitored by flights of the space shuttle and by SATELLITES. Degenerate rain forest is JUNGLE, although the two terms are often confused.

Rain gauge. Instrument for measuring RAINFALL, usually consisting of a cylindrical container open to the sky.

Rain shadow. Area of low PRECIPITATION located on the LEEWARD side of a topographic barrier such as a mountain RANGE. Moisture-laden WINDS are forced to rise, so they cool ADIABATICALLY, leading to CONDENSATION and precipitation on the WINDWARD side of the barrier. When the AIR descends on the other side of the MOUNTAIN, it is dry and relatively warm. The area to the east of the Rocky Mountains is in a rain shadow.

Rainfall. The amount of water a PLACE receives from the ATMOSPHERE over a given period. Meteorologists describe light rainfall as 0.1 inch

(2.5 millimeters) in one hour; moderate rainfall as 0.1 to 0.3 inch (2.5-7.6 millimeters) in one hour; and heavy rainfall as more than 0.3 inch (7.6 millimeters) in one hour. Climatologists are concerned with the average annual rainfall, the average over more than twenty years of rainfall records. The place with the highest annual average rainfall is Mount Waialeale, on the ISLAND of Hawaii, which receives on average 460 inches (11,700 millimeters) per year. The record for the highest rainfall in a single year belongs to Cherrapunji, in the foothills of the Himalayas, which received 1,042 inches (26,467 millimeters) in the year 1860-1861. Rainfall is measured using a RAIN GAUGE. Rainfall should not be confused with PRECIPITATION, which includes SNOW, SLEET, and hail.

Range. Difference between the highest and lowest values in a record. The TEMPERATURE range can be measured daily as the difference between the maximum and minimum temperatures, or calculated as a monthly or annual value. The TIDAL RANGE is the difference in height between the height of water at the extremes of high and low tides. This can be measured daily, or calculated for monthly or annual figures.

Range, mountain. Linear series of MOUNTAINS close together, formed in an OROGENY, or mountain-building episode. Tall mountain ranges such as the Rocky Mountains are geologically much younger than older mountain ranges such as the Appalachians.

Mountain ranges are series of mountains close together, formed in an orogeny, or mountain-building episode. (PhotoDisc)

Rank-size rule. Relationship between the POPULATION size of a CITY and its place in a hierarchy of cities within a COUNTRY. The rule is expressed mathematically—the nth largest city in a REGION is $1/n$ times the size of the largest city in the region. In other words, if the largest city has a population of 1 million, the fifth-largest city would have a population of one-fifth of that, or 200,000 people. The URBAN population of the United States fits the rank-size rule. The rank-size DISTRIBUTION is distorted by the presence in a country of a PRIMATE CITY, one that is disproportionally large. Mexico City, for example, is about ten times larger than Mexico's next-largest city, Guadalajara.

Rapids. Stretches of RIVERS where the water flow is swift and turbulent because of a steep and rocky CHANNEL. The turbulent conditions are called WHITE WATER. If the change in ELEVATION is greater, as for small WATERFALLS, they are called CATARACTS.

Rapids are stretches of rivers in which water flow is swift and turbulent through steep and rocky channels. (PhotoDisc)

Rational horizon. See HORIZON, TRUE.

Realm. In its older and narrowest sense, a realm is a kingdom. In its broader geographical sense, it is any political territory.

Recessional moraine. Type of TERMINAL MORAINE that marks a position of shrinkage or wasting or a GLACIER. Continued forward flow of ice is maintained so that the debris that forms the moraine continues to accumulate. Recessional moraines occur behind the terminal moraine.

Recumbent fold. Overturned fold in which the upper part of the fold is almost horizontal, lying on top of the nearest adjacent surface.

Reef (geology). VEIN of ORE, for example, a reef of gold.

Reef (marine). Underwater ridge made up of sand, rocks, or coral that rises near to the water's surface. See also CORAL REEF.

Refraction of waves. Bending of waves, which can occur in all kinds of waves. When OCEAN WAVES approach a COAST, they start to break as they approach the SHORE because the depth decreases. The wave speed is retarded and the WAVE CREST seems to bend as the wavelength decreases. If waves are approaching a coast at an oblique angle, the crest line bends near the shore until it is almost parallel. If waves are approaching a BAY, the crests are refracted to fit the curve of the bay.

Region. Area of the earth that is homogeneous in respect to certain chosen characteristics. Geographers tend to use multiple features or criteria to identify regions. CLIMATE, TOPOGRAPHY, and LANDFORMS might be used to differentiate physical regions. LANGUAGE, ethnicity, and CULTURE can be used to distinguish human regions. Industrial regions are based on their production, for example, the COTTON BELT. Regional geographers divide the world into FORMAL REGIONS, those with a measurable and usually visible homogeneity. Other regions might be defined as FUNCTIONAL REGIONS, having a definite CORE or node, such as a CITY and its HINTERLAND.

Regionalism. Feeling of collective identity by the people of a REGION, based on their personal identification with that region. Texans, for example, often display regionalism. Regionalism can influence ETHNIC GROUPS, whose aims may include increased political power or autonomy. The Basque people of the region on the west BORDER of France and Spain have created a strong regional movement, whose adherents use violent methods in their struggle for a Basque HOMELAND. The region where these feelings are expressed is a VERNACULAR REGION.

Regolith. Layer of broken ROCK at the earth's surface, lying over BEDROCK. Over time, regoliths can weather further and break down into SOIL. Regolith comprises the C horizon of a SOIL PROFILE. Slopes below CLIFFS carry a layer of regolith.

Regression. Retreat of the SEA from the land; it allows land EROSION to occur on material formerly below the sea surface.

Rejuvenation. STREAM or LANDSCAPE is rejuvenated when there is an increase in RELIEF, generally because of tectonic UPLIFT of the surface. This puts new kinetic ENERGY into the system, creating a new, lower BASE LEVEL. EROSION occurs more rapidly. LANDFORMS such as KNICKPOINTS and incised MEANDERS are evidence of rejuvenation.

Relative humidity. Measure of the HUMIDITY, or amount of moisture, in the ATMOSPHERE at any time and place compared with the total amount of moisture that same AIR could theoretically hold at that TEMPERATURE. Relative humidity is a ratio that is expressed as a percentage. When the air is saturated, the relative humidity reaches 100 percent and rain occurs. When there is little moisture in the air, the relative humidity is low, perhaps 20 percent. Relative humidity varies inversely with temperature, because warm air can hold more moisture than cooler air. Therefore, when temperatures fall overnight, the air often becomes saturated and DEW appears on grass and other surfaces. The human COMFORT INDEX is related to the relative humidity. Hot temperatures are more bearable when relative humidity is low. Media announcers frequently use the term "humidity" when they mean relative humidity.

Relative location. The location of one PLACE in relation to another place, for example, "west of the Mississippi."

Relief. In a LANDSCAPE, the difference in ELEVATION between the highest and lowest points. MOUNTAINS cut by STREAMS are areas of high relief; PLAINS are areas of low relief, although they may be at quite high elevation. A PLATEAU is a feature of high elevation but low relief.

Religion. System of belief in gods, spirits, or sacred objects. The major religions of the world are Christianity, ISLAM, Buddhism, Hinduism, and Judaism, but there are hundreds of others, as well as many branches and denominations within the major religions. Religions combine a belief system and worship with moral behavior, as well as ceremonies and institutions. Religion is an important aspect of CULTURE.

Remote sensing. Gathering information about the earth from some distance. Aerial photography and SATELLITE imagery are widely used forms of remote sensing, allowing scientists to learn much about PLACES without having to visit them in person.

Renewable resource. Renewable resources are generally living RESOURCES that can be grown and replaced; however, INSOLATION, or sunlight, is also considered an important renewable resource.

Replacement rate. The rate at which females must reproduce to maintain the size of the POPULATION. It corresponds to a FERTILITY RATE of 2.1.

Republic. System of government in which supreme power is held by representatives elected by members of the public. A republic cannot be a MONARCHY.

Reservoir. Artificial LAKE in which water is stored, for example, for IRRIGATION or for watering animals.

Reservoir rock. Geologic ROCK layer in which OIL and gas often accumulate; often SANDSTONE or LIMESTONE.

Resource. Something useful, for example, materials, services, or information. Earth scientists are often concerned with natural resources, or goods and services supplied by the natural ENVIRONMENT, as opposed to human resources, such as experience, wisdom, skill, or labor. Natural resources are generally classified as RENEWABLE and NONRENEWABLE RESOURCES. Renewable or living resources include FORESTS, plants such as grains and fruits, animals, and fish. Nonrenewable or nonliving resources include MINERALS and fuels. Humans also appreciate intangible resources, such as open space, personal satisfaction, beauty, and other abstractions. These nonmaterial resources can be economically important. Resources that are held in common, such as the OCEANS or the ATMOSPHERE, are the hardest to protect, because each individual believes his or her actions have little impact.

Respiration. Metabolic process found in animals and microbes whereby complex organic molecules (food) are oxidized to CARBON DIOXIDE, thus releasing ENERGY for work.

Retrograde orbit. ORBIT of a SATELLITE around a PLANET that is in the opposite sense (direction) in which the planet rotates.

Retrograde rotation. ROTATION of a PLANET in a direction opposite to that of its REVOLUTION.

Reverse fault. Feature produced by compression of the earth's CRUST, leading to crustal shortening. The UPTHROWN BLOCK overhangs the downthrown block, producing a FAULT SCARP where the overhang is prone to LANDSLIDES. When the movement is mostly horizontal, along a low angle FAULT, an overthrust fault is formed. This is commonly associated with extreme FOLDING.

High-angle reverse fault in Woburn, Quebec. (Geological Survey of Canada)

Reverse polarity. Orientation of the earth's MAGNETIC FIELD so that a COMPASS needle points to the SOUTHERN HEMISPHERE.

Revolution. The annual movement of the earth around the SUN in an elliptical ORBIT that takes 365.2422 DAYS.

Ria coast. Ria is a long narrow ESTUARY or RIVER MOUTH. COASTS where there are many rias show the effects of SUBMERGENCE of the land, with the SEA now occupying former RIVER VALLEYS. Generally, there are MOUNTAINS running at an angle to the coast, with river valleys between each RANGE, so that the ria coast is a succession of estuaries and promontories. The submergence can result from a rising SEA LEVEL, which is common since the melting of the PLEISTOCENE GLACIERS, or it can be the result of SUBSIDENCE of the land. There is often a great TIDAL RANGE in rias, and in some, a tidal BORE occurs with each TIDE. The eastern coast of the United States, from New York to South Carolina, is a ria coast. The southwest coast of Ireland is another. The name comes from Spain, where rias occur in the south.

Richter scale. SCALE used to measure the magnitude of EARTHQUAKES; named after American physicist Charles Richter, who, together with Beno Gutenberg, developed the scale in 1935. The scale is a quantitative measure that replaced the older MERCALLI SCALE, which was a descriptive scale. Numbers range from zero to nine, although there is no upper limit. Each whole number increase represents an order of magnitude, or an increase by a factor of ten. The actual MEASUREMENT was logarithm to base 10 of the maximum SEISMIC WAVE amplitude (in thousandths of a millimeter) recorded on a standard SEISMOGRAPH at a distance of 60 miles (100 km.) from the earthquake EPICENTER.

Ridge. Long narrow LANDFORM of high ELEVATION. The top or crest of a ridge is the ridgeline, but this is often referred to simply as a ridge. In forested REGIONS, logging roads are often constructed along ridgelines.

Rift. Portion of the earth's CRUST where TENSION has caused faulting, producing an elongate BASIN; rifts fill with SEDIMENTS and, sometimes, VOLCANIC ROCKS.

Rift propagation. Lateral movement of a rifting process that leads to the prying open of a section of the LITHOSPHERE, accompanied by the formation of IGNEOUS ROCKS.

Rift valley. Long, low REGION of the earth's surface; a VALLEY or TROUGH with FAULTS on either side. Unlike valleys produced by EROSION, rift valleys are produced by tectonic forces that have caused the faults or fractures to develop in the ROCKS of Earth's CRUST. TENSION can lead to the block of land between two faults dropping in ELEVATION compared to the surrounding blocks, thus forming the rift valley. A small LANDFORM produced in this way is called a GRABEN. A rift valley is a much larger feature. In Africa, the Great Rift Valley is partially occupied by Lake Malawi and Lake Tanganyika, as well as by the Red Sea.

Rills. Small trickles of water in a CATCHMENT area or WATERSHED. They form and enlarge through EROSION, eventually joining to form gullies.

Ring dike. Volcanic LANDFORM created when MAGMA is intruded into a series of concentric FAULTS. Later EROSION of the surrounding material may reveal the ring dike as a vertical feature of thick BASALT rising above the surroundings.

Ring of Fire. Zone of volcanic activity and associated EARTHQUAKES that marks the edges of various TECTONIC PLATES around the Pacific Ocean, especially those where SUBDUCTION is occurring.

Riparian. Term meaning related to the BANKS of a STREAM or RIVER. Riparian VEGETATION is generally trees, because of the availability of moisture.

Riparian rights. Legal regulations that allow the use of water in a STREAM by anyone who owns land through which the stream flows, provided that they do not prevent the water from continuing its downstream flow. In the United States, laws regarding riparian rights are controversial and vary from STATE to state.

River. Naturally occurring STREAM of water flowing in a natural CHANNEL. Many earth scientists, including geographers, prefer the term stream. Rivers that always contain flowing water are PERENNIAL STREAMS. Those that flow only for part of the year are INTERMITTENT STREAMS. A watercourse that has water for only a DAY or so is called an EPHEMERAL STREAM. An EXOTIC STREAM, or river, is one that flows through a DESERT, receiving its waters from some distant REGION.

Both "river" and "stream" are used to apply to a body of water that flows in a natural channel. (PhotoDisc)

River, mouth of. See MOUTH OF RIVER.

River terraces. LANDFORMS created when a RIVER first produces a FLOOD-PLAIN, by DEPOSITION of ALLUVIUM over a wide area, and then begins downcutting into that alluvium toward a lower BASE LEVEL. The re-newed EROSION is generally because of a fall in SEA LEVEL, but can result from tectonic UPLIFT or a change in CLIMATE pattern due to increased PRECIPITATION. On either side of the river, there is a step up from the new VALLEY to the former alluvium-covered floodplain surface, which is now one of a pair of river terraces. This process may occur more than once, creating as many as three sets of terraces. These are called depositional terraces, because the terrace is cut into river deposits. Ero-sional terraces, in contrast, are formed by lateral migration of a river, from one part of the valley to another, as the river creates a floodplain. These terraces are cut into BEDROCK, with only a thin layer of alluvium from the point BAR deposits, and they do not occur in matching pairs.

River valleys. VALLEYS in which STREAMS flow are produced by those streams through long-term EROSION and DEPOSITION. The LANDFORMS produced by FLUVIAL action are quite diverse, ranging from spectacu-lar CANYONS to wide, gently sloping valleys. The patterns formed by stream networks are complex and generally reflect the BEDROCK geol-ogy and TERRAIN characteristics.

Roadstead. Coastal anchorage for ships lacking the protection of a HAR-BOR.

Nineteenth century roadstead harbor of Corinto, on Nicaragua's Pacific coast. (Arkent Archive)

Roches moutonnées. Erosional feature formed usually by continental GLACIATION. As an ICE SHEET advanced over a piece of resistant BED-ROCK, it polished and smoothed the front side, while on the lee side PLUCKING removed sections, leading to a jagged profile. The name "rock sheep" is thought to indicate a resemblance to a sheep lying down; another explanation is that men in the nineteenth century wore wigs made of sheepskin, with the wool attached. These ROCKS look like wigs lying on a flat surface; the front is smooth, but the back is rough and curly. In English, the term "sheep rock" is sometimes used.

Rock. Naturally occurring combination of MINERALS, which make up the earth. Rocks are divided into three classes. IGNEOUS ROCKS are formed when MAGMA, or molten material, cools and solidifies. SEDIMENTARY ROCKS are formed when fragments of other rocks are cemented together. METAMORPHIC ROCKS are those that have been changed by heat and pressure; they originally may have been igneous or sedimentary.

Rock avalanche. Extreme case of a rockfall. It occurs when a large mass of ROCK moves rapidly down a steeply sloping surface, taking everything that lies in its path. It can be started by an EARTHQUAKE, rock-blasting operations, or vibrations from thunder or artillery fire.

Rock cycle. Cycle by which ROCKS are formed and reformed, changing from one type to another over long PERIODS of geologic time. IGNEOUS ROCKS are formed by cooling from molten MAGMA. Once exposed at the surface, they are subject to WEATHERING and EROSION. The products of erosion are compacted and cemented to form SEDIMENTARY ROCKS. The heat and pressure accompanying a volcanic intrusion causes adjacent rocks to be altered into METAMORPHIC ROCKS.

Rock fall. Rapid fall of blocks of ROCK. It is often the result of FROST WEDGING of rocks in an exposed CLIFF in a mountainous REGION. EARTHQUAKES can also cause rock falls.

Rock flour. Fine, powderlike material at the base of a GLACIER, produced by the constant ABRASION as the glacier and its MORAINE grind along.

Rock glacier. Form of mass movement in high MOUNTAINS that produces a lobe or tongue of broken ROCKS that moves slowly down a VALLEY. The rock is SCREE produced by FROST WEDGING. PRECIPITATION, or MELTWATER, forms an adjacent glacier and penetrates between the debris, freezing into an ice mass in the central portion. The whole mass then moves slowly downhill like a glacier. Also called a stone RIVER.

Rock salt. Sodium chloride in crystalline form. It is formed by the EVAPORATION of seawater and can be found in underground layers or beds of great thickness. Such deposits indicate the gradual evaporation of an enclosed SEA. In arid REGIONS, rock salt is obtained at the surface, through evaporation in shallow SALINE LAKES. Rock salt is mined for both domestic use and industrial purposes. In previous centuries, the need to obtain salt by mining made it an expensive commodity in Europe. SALT DOMES are an important source of rock salt. Called halite by mineralogists.

Rock slide. Event that occurs when water lubricates an unconsolidated mass of weathered ROCK on a steep slope, causing rapid downslope movement. In a RIVER VALLEY where there are steep SCREE slopes being constantly carried away by a swiftly flowing STREAM, the undercutting at the base can lead to constant rockslides of the surface layer of rock. A large rockslide is a ROCK AVALANCHE.

Ropy lava. Extremely viscous LAVA; a hot basaltic flow that cools and hardens into smooth to ropy surfaces, displaying clearly the flow lines. In Hawaii, called pahoehoe.

Lava is the magma, or molten material from within the earth, that emerges at the surface. (PhotoDisc)

Rotation. Turning of the earth on its AXIS, in an eastward, or COUNTER-CLOCKWISE, direction, once every 23 hours, 56 minutes, and 4 seconds. Rotation affects the WINDS, OCEAN CURRENTS, TIDES, and length of DAY. Because of the rotation, the SUN appears to travel from east to west each day, although its position in our SOLAR SYSTEM is fixed. The rotational velocity of any point on Earth varies with its LATITUDE; a point at the EQUATOR travels at more than 1,000 miles (1,600 km.) an hour while the velocity at the POLES is zero.

Runoff. Water that becomes part of a STREAM. Water generally comes to the stream through PRECIPITATION moving as sheetflow over the land surface, or in CHANNELS such as RILLS and gullies. Also can include GROUNDWATER that flows into the stream. Total runoff is less than total precipitation for any WATERSHED, because some water is lost through EVAPORATION and some water enters the groundwater storage.

Rural. Society or SETTLEMENT in which there are a small number of inhabitants in a large area of land. AGRICULTURE is the typical economic sector. Governments classify their POPULATIONS into URBAN or rural, but the precise definition varies greatly from one COUNTRY to another. In the United States, a PLACE with more than twenty-five hundred residents is considered urban; in Japan, the urban category starts at thirty thousand.

S waves. Type of SEISMIC disturbance of the earth when an EARTHQUAKE occurs. In an S wave, particles move about at right angles to the direction in which the wave is traveling. S waves cannot pass through the earth's CORE, which is why scientists believe the INNER CORE is liquid. Also called transverse wave, shear wave, or secondary wave.

Sacred space. SITE or area recognized by certain religious groups as deserving special attention because of its connection with religious figures or religious events. Sacred sites or spaces are maintained by believers over many centuries. Pilgrims make journeys to sacred sites or spaces. The *hajj* is the pilgrimage to the sacred PLACE of Mecca that all Muslims must try to make once in their lifetime. Lourdes in France is a sacred space visited by hundreds of thousands of Roman Catholic pilgrims each year.

Saddle. See COL.

Sahel. Southern edge of the Sahara Desert; a great stretch of semiarid land extending from the Atlantic Ocean in Senegal and Mauritania through Mali, Burkina Faso, Nigeria, Niger, Chad, and Sudan. Northern Ethiopia, Eritrea, Djibouti, and Somalia usually are included also. This transition zone between the hot DESERT and the tropical SAVANNA has low summer RAINFALL of less than 8 inches (200 millimeters) and a natural VEGETATION of low grasses with some small shrubs. The REGION traditionally has been used for PASTORALISM, raising goats, camels, and occasionally sheep. Since a prolonged DROUGHT in the 1970's, DESERTification, soil EROSION, and FAMINE have plagued the Sahel. The narrow band between the northern Sahara and the Mediterranean North African COAST is also called Sahel. "Sahel" is the Arabic word for edge.

Salina. See ALKALI FLAT.

Saline lake. LAKE with elevated levels of dissolved solids, primarily resulting from evaporative concentration of SALTS; saline lakes lack an out-

California's Mono Lake is an interesting example of a naturally saline lake whose chemical salt concentrations have risen even higher because of human intervention. (PhotoDisc)

let to the SEA. Well-known examples include Utah's Great Salt Lake, California's Mono Lake and Salton Sea, and the Dead Sea in the Middle East.

Salinity. Measure of the concentration of dissolved SALTS in seawater. Salinity is the amount of salt in grams dissolved in one kilogram of seawater. The value is written in parts per thousand. Average salinity of the OCEANS is 35 parts per thousand.

Salinization. Accumulation of SALT in SOIL. When IRRIGATION is used to grow crops in semiarid to arid REGIONS, salinization is frequently a problem. Because EVAPORATION is high, water is drawn upward through the soil, depositing dissolved salts at or near the surface. Over years, salinization can build up until the soil is no longer suitable for AGRICULTURE. The solution is to maintain a plentiful flow of water while ensuring that the water flows through the soil and is drained away.

Salt. In chemistry, a substance formed when an acid reacts with a base. In everyday terms, salt refers to sodium chloride or table salt, which is the most common form of salt.

Salt domes. Formations created when deeply buried salt layers are forced upwards. SALT under pressure is a plastic material, one that can flow or move slowly upward, because it is lighter than surrounding SEDIMENTARY ROCKS. The salt forms into a plug more than a half mile (1 km.) wide and as much as 5 miles (8 km.) deep, which passes through overlying sedimentary rock layers, pushing them up into a dome shape as it passes. Some salt domes emerge at the earth's surface; others are close to the surface and are easy to mine for ROCK SALT. OIL and NATURAL GAS often accumulate against the walls of a salt dome. Salt domes are numerous around the COAST of the Gulf of Mexico, in the North Sea REGION, and in Iran and Iraq, all of which are major oil-producing regions.

Salt water. Water with a SALT content of 3.5 percent, such as is found in normal OCEAN water.

Saltation. Process whereby a particle is moved forward by water or WIND, being lifted, carried, and then dropped, over and over. It comes from the Latin word for "jump" and has nothing to do with table SALT.

Saltwater intrusion. AQUIFER contamination by salty waters that have migrated from deeper aquifers or from the SEA.

Saltwater lake. See SALINE LAKE.

Saltwater wedge. Wedge-shaped intrusion of seawater from the OCEAN into the bottom of a RIVER; the thin end points upstream.

Sand. Grain or particle size with a diameter ranging between 0.0008 inch (0.02 millimeter) for fine sand to 0.08 inch (2 millimeters) for coarse sand. Sand can be many colors. Most sand is composed mostly of QUARTZ and is formed by EROSION of granitic ROCKS. The resulting quartz sand is yellowish. BASALT weathers into black sand. Coral forms white or occasionally pink sand.

Sand dunes. Accumulations of SAND in the shape of mounds or RIDGES. They occur on some COASTS and in arid REGIONS. Coastal dunes are formed when the prevailing WINDS blow strongly onshore, piling up sand into dunes, which may become stabilized when grasses grow on them. DESERT sand dunes are a product of DEFLATION, or wind ERO-SION removing fine materials to leave a DESERT PAVEMENT in one region and sand deposits in another. Sand dunes are classified by their shape into barchans, or crescent-shaped dunes; seifs or LONGITUDINAL DUNES; TRANSVERSE DUNES; star dunes; and sand drifts or sand sheets.

Sand spit. See SPIT.

Sandbar. When BEACH SAND is moved by WAVES and LONGSHORE CUR-RENTS, it can form long narrow deposits called BARS. They are named according to their position. An offshore bar is parallel to the COAST. A baymouth bar encloses a BAY, running from one HEADLAND to the other. A bar extending outward from the land at one end is a SPIT; a connecting bar is known as a TOMBOLO. Bars are unstable and tempo-rary LANDSCAPE features. Along the southern and eastern coasts of the United States are huge sandbars called BARRIER ISLANDS.

Red sandstone formation near Arbroath, Scotland. Historically, this sandstone was used both for building and for ballast in ships. (Ray Sumner)

Sandstone. Common SEDIMENTARY ROCK produced through the LITHIFI-CATION of sand-sized grains. The pore spaces between the grains may be empty, filled with AIR, or filled with a cementing material such as calcium carbonate.

Sapping. Natural process of EROSION at the bases of HILL slopes or CLIFFS whereby support is removed by undercutting, thereby allowing overly-ing layers to collapse; SPRING SAPPING is the facilitation of this process by concentrated GROUNDWATER flow, generally at the heads of VALLEYS.

Satellite. Small object that revolves around a larger object. The Moon is a natural satellite of Earth. In this SOLAR SYSTEM, only Mercury and Venus have no satellites. Humans have also created many artificial satellites, the first of these being Sputnik 1, launched on October 4, 1957, by the Soviet Union. Hundreds of satellites now ORBIT Earth. Satellites are used for COMMUNICATIONS, military purposes, and scientific research, such as WEATHER FORECASTING and studying VEGETATION, OCEANS, and atmospheric changes. The GLOBAL POSITIONING SYSTEM (GPS) uses signals from satellites to accurately obtain ABSOLUTE LOCATIONS on Earth. This military application was developed by the United States in the late 1970's and made available to the public in the 1990's. Geographers make wide use of imagery from the series of satellites named LANDSAT (also Earth Resources Technology Satellite) and a series of satellites named GOES (Geostationary Operational Environmental Satellite). This type of research is called REMOTE SENSING.

Satellite meteorology. Study of atmospheric phenomena using SATELLITE data; an indispensable tool for forecasting WEATHER and studying CLIMATE on a global scale.

Saturation, zone of. Underground REGION below the zone of AERATION, where all pore space is filled with water. This water is called GROUNDWATER; the upper surface of the zone of saturation is the WATER TABLE.

Savanna. VEGETATION that consists of tall grass with occasional trees and shrubs interspersed. Savanna occurs in the TROPICS, between the tropical RAIN FOREST and the semiarid REGIONS that fringe true DESERTS. The CLIMATE is tropical with rain concentrated in the summer months, followed by a long dry season. In some countries, people divide the climate of these areas into two SEASONS—wet and dry—because in the Tropics, TEMPERATURES are high all year. Because the rain falls in summer, much moisture is lost through EVAPORATION, so moisture conditions are insufficient for FOREST growth. Trees are scarce and small to medium in height, with small leaves, spreading crowns, and an extensive root system. SOILS of the savanna are more fertile than those of the tropical rain forest, and some farming is undertaken in these areas, especially if water for IRRIGATION is available. Savannas cover about 40 percent of the earth's lands. They are thought to have been extended through the human practice of setting fire to the dry grasses at the end of the dry season in order to ensure fresh new growth when the rains came; therefore, many plants of the savanna are fire tolerant. The savanna BIOME is particularly extensive in Africa. In South America, the savanna of Venezuela is called LLANOS; it is called Campo Cerrado or Pantanal in different parts of Brazil. Savanna also occurs in India, Madagascar, and Thailand. The name is sometimes spelled savannah. See also GRASSLAND.

Scale. Relationship between a distance on a MAP or diagram and the same distance on the earth. Scale can be represented in three ways. A linear, or graphic, scale uses a straight line, marked off in equally spaced intervals, to show how much of the map represents a mile or a kilometer. A representative fraction (RF) gives this scale as a ratio. A verbal scale uses words to explain the relationship between map size and actual size. For example, the RF 1:63,360 is the same as saying "one inch to the mile."

Scarp. Short version of the word "ESCARPMENT," a short steep slope, as at the edge of a PLATEAU. EARTHQUAKES lead to the formation of FAULT SCARPS.

Schist. METAMORPHIC ROCK that can be split easily into layers. Schist is commonly produced from the action of heat and pressure on SHALE or SLATE. The rock looks flaky in appearance. Mica-schists are shiny because of the development of visible mica. Other schists include talc-schist, which contains a large amount of talc, and hornblende-schist, which develops from basaltic rocks.

Scree. Broken, loose ROCK material at the base of a slope or CLIFF. It is often the result of FROST WEDGING of BEDROCK cliffs, causing rockfall. Another name for scree is TALUS.

Sea. Part of an OCEAN that is partially enclosed by land. Seas occur at the margins of oceans. Well-known seas include the Caribbean, Mediterranean, Red, Black, and North. There is no clear distinction in naming water features, however. For example, the Bay of Bengal might be termed a sea. On the other hand, some saltwater LAKES are misnamed seas. Examples are the Caspian Sea, Aral Sea, and Dead Sea. These are all lakes, because they are totally landlocked and are not part of a larger ocean.

Sea fog. See ADVECTION FOG.

Sea level. Standard reference height, which is used as a basis for all ELEVATIONS above or below for terrestrial or submarine elevations, respectively. The height of the sea-land interface is constantly changing, mainly because of the EBB and flow of TIDES. CURRENTS, WINDS, pressure conditions, and other factors also have an effect. When an elevation is given as a height above sea level, this refers to a height above MEAN SEA LEVEL. Mean sea level (MSL) is calculated from average hourly tidal records over many years. For the United States, records are assembled for more than forty tidal gauges, together with data from the TOPEX Poseidon SATELLITE. The MSL of the Gulf of Mexico is higher than the MSL of the Atlantic COAST. Florida has the lowest MSL in the United States; Oregon has the highest. Sea levels have changed throughout the earth's history. Since the last ICE AGE, around fifteen thousand years ago, sea level has risen because of the melting ice. On average, the increase in sea level is about 400 feet (130 meters). The prediction that GLOBAL WARMING will cause a rise in sea level in the near future is a cause of concern, because so many people throughout the world live close to the coast. A rise of only 1 foot (0.3

meter) would destroy billions of dollars worth of valuable real estate, inundate rich farmlands, and completely cover the HOMELANDS of some ISLAND NATIONS in the Pacific and Indian Oceans.

Sea lane. See SEAWAY.

Seafloor spreading. Term often used to refer to the separation of the OCEAN floor at a spreading center located along a MID-OCEAN RIDGE. The theory was advanced in the 1960's, and new evidence over the following decades confirmed the hypothesis. Seafloor spreading occurs where TECTONIC PLATES are diverging, or moving apart, and new CRUST is being created. Volcanic ERUPTIONS, fractures, and EARTHQUAKES accompany seafloor spreading. The spreading is balanced by SUBDUCTION, when plates converge and crust is destroyed.

Seamount. Large VOLCANO rising more than 3,000 feet (1,000 km.) from the OCEAN floor to near the surface. LAVA erupts from a fracture or RIFT on the ocean floor. Oceanographic research has shown that there are twenty thousand seamounts in the world oceans. A seamount with a flat top is called a GUYOT; these features are important to the explanation of CORAL REEFS and ATOLLS throughout the Pacific Ocean.

Seasons. An Earth year is conventionally divided into four seasons— spring, summer, autumn or fall, and winter. The division into seasons is based on TEMPERATURE changes, which are related to changes in the length of DAY. This is caused by the tilt of the earth's AXIS at 23.5 DEGREES from vertical, which means that the CIRCLE OF ILLUMINATION changes in the course of a year as the earth revolves around the SUN. In North America, the seasons are said to start at the summer SOLSTICE, the autumnal EQUINOX, the WINTER SOLSTICE, and the VERNAL or spring equinox. In the SOUTHERN HEMISPHERE, the pattern of seasons is reversed, so that summer in the NORTHERN HEMISPHERE corresponds to winter in the Southern Hemisphere. In polar REGIONS, seasons are extreme, with darkness throughout the winter and daylight throughout the summer. At the EQUATOR, the opposite holds; there is no variation in daylength throughout the year. Countries that experience the MONSOON, especially India and Southeast Asia, experience only three seasons—hot-wet, cool-dry, and hot-dry.

Seaway. Route traveled by ships on the open SEAS. Often the word is used to refer to the actual passage by the ship. In large seas, a ship makes a heavy seaway. A seaway is also a large CANAL constructed to provide interior access for large oceangoing ships, such as the Saint Lawrence Seaway, which connects North America's Great Lakes to the Atlantic Ocean by way of the Gulf of Saint Lawrence. Also called a sea lane.

Secondary industry. That part of the ECONOMY that takes the raw materials produced by the primary sector of the economy and processes them into salable products. At a simple level, this might be the milling of grain into flour, or the sawing of logs into lumber. Secondary industry also involves heavy and technically sophisticated industries such as steel production, shipbuilding, and automobile manufacture. Some economists distinguish between light industry and heavy industry.

Secondary waves. See S WAVES.

Sectionalism. Form of extreme devotion to local interests and customs.

Sediment. Solid earth material that has been weathered and is deposited after being transported by water, ice, or WIND, or moved downward by GRAVITY. Sedimentation refers to the laying down of deposits that, after consolidation or cementation, become SEDIMENTARY ROCKS.

Light-colored sediment from the Mississippi River flowing into the Gulf of Mexico. (U.S. Geological Survey)

Sedimentary rocks. ROCKS formed from SEDIMENTS that are compressed and cemented together in a process called LITHIFICATION. Sedimentary rocks cover two-thirds of the earth's land surface but are only a small proportion of the earth's CRUST. SANDSTONE is a common sedimentary rock. Sedimentary rocks form STRATA, or layers, and sometimes contain FOSSILS.

Segregation. Spatial separation of a subgroup of a POPULATION, often because of discrimination. Ethnic ENCLAVES are a largely voluntary form of segregation, because the residents have a close sense of community and internal cohesion. An extreme example of such segregation is called a ghetto. Apartheid was a form of segregation in South Africa based on racial discrimination, the separation of white and non-white—largely black—populations.

Seif dunes. Long, narrow RIDGES of SAND, built up by WINDS blowing at different times of year from two different directions. Seif dunes occur in parallel lines of sand over large areas, running for hundreds of miles in the Sahara, Iran, and central Australia. Another name for seif dunes is LONGITUDINAL DUNES. The Arabic word means sword.

Seismic. Pertaining to EARTHQUAKES.

Seismic activity. Movements within the earth's CRUST that often cause various other geological phenomena to occur; the activity is measured by SEISMOGRAPHS.

Seismic belt. REGION of relatively high SEISMICITY, globally distributed; seismic BELTS mark regions of PLATE interactions.

Seismic gap. FAULT REGION known to have had previous EARTHQUAKES but not within the area's most recent recurrence period.

Seismic wave. TSUNAMI or an OCEAN WAVE caused by a seismic event under the ocean.

Seismicity. Occurrence of EARTHQUAKES, which is expressed as a function of LOCATION and time.

Seismogram. Image of EARTHQUAKE wave vibrations recorded on paper, photographic film, or a video screen.

Seismograph. Instrument used to record the ground shaking that occurs with an EARTHQUAKE. In a simple seismograph, a paper is attached to a

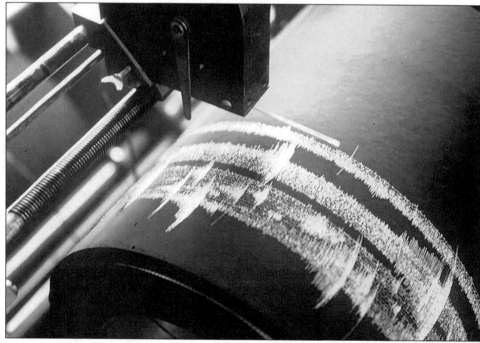

Smoke-drum seismographic record. (U.S. Geological Survey)

rotating drum, and a pen is attached to an arm that is firmly embedded in the ground, so that it vibrates when the earth moves. Seismographs also record earth movements caused by atomic explosions.

Seismology. The scientific study of EARTHQUAKES. It is a branch of GEOPHYSICS. The study of SEISMIC WAVES has provided a great deal of knowledge about the composition of the earth's interior.

Seismometer. Instrument that measures the motion of the ground, used to record SEISMIC ENERGY; also known as a geophone or a seismic detector.

Self-determination. Right of a group of people who occupy a distinct territory to control that territory and determine their own future development or destiny. The Palestinians have fought both physically and politically for the right to control their own territory, or STATE, and thus to enjoy the benefits of self-determination.

Semidesert. REGION with DESERT characteristics but with greater PRECIPITATION than a true desert.

Service sector. See TERTIARY INDUSTRY.

Settlement. Small community of people and their residences. A settlement is smaller than a TOWN, so settlements are generally found in RURAL areas. Cultural geographers study settlement patterns. Clustered settlements, where the houses are relatively close together, are common in Europe and Asia; dispersed settlements are more common in rural parts of North America, in Australia, and in Africa.

Shadow zone. When an EARTHQUAKE occurs at one LOCATION, its waves travel through the earth and are detected by SEISMOGRAPHS around the world. Every earthquake has a shadow zone, a band where neither P nor S WAVES from the earthquake will be detected. This shadow zone leads scientists to draw conclusions about the size, density, and composition of the earth's CORE.

Shale. SEDIMENTARY ROCK consisting of layers of fine-grained materials of CLAY or SILT size. Shale is the most abundant of the sedimentary rocks. It is a raw material for brick making and ceramics.

Shale oil. SEDIMENTARY ROCK containing sufficient amounts of hydrocarbons that can be extracted by slow distillation to yield OIL.

Shallow-focus earthquakes. EARTHQUAKES having a focus less than 35 miles (60 km.) below the surface.

Shantytown. URBAN SQUATTER SETTLEMENT, usually housing poor newcomers.

Shear waves. See S WAVES.

Sheet erosion. See SHEET WASH.

Sheet wash. When water flows as a thin sheet across a slope, it can erode loose materials such as SOIL particles that previously were dislodged by SPLASH EROSION. This generally occurs after a sudden and intense period of PRECIPITATION. It can lead to considerable loss of TOPSOIL on an unplanted field. Sheet wash can also occur on BEDROCK surfaces. In nature, sheet wash is less common than channeled flow, where the rainwater gathers into RILLS; in URBAN AREAS, however, sheet wash can be seen on streets during a rainstorm as water flows toward storm drains. Also called sheet erosion.

Sheikdom. Islamic COUNTRY whose ruler bears the title of *sheik*.

Shelter belt. Another word for a WINDBREAK.

Shield. Large part of the earth's CONTINENTAL CRUST, comprising very old ROCKS that have been eroded to REGIONS of low RELIEF. Each CONTINENT has a shield area. In North America, the Canadian Shield extends from north of the Great Lakes to the Arctic Ocean. Sometimes known as a CONTINENTAL SHIELD.

Shield volcano. VOLCANO created when the LAVA is quite viscous or fluid and highly basaltic. Such lava spreads out in a thin sheet of great radius

The Hawaiian Islands contain some of the greatest shield volcanoes on earth. Several of those volcanoes can be seen in this 1988 satellite photograph of Maui (left) and Hawaii (right). (Corbis)

but comparatively low height. As flows continue to build up the volcano, a low DOME shape is created. The greatest shield volcanoes on Earth are the ISLANDS of Hawaii, which rise to a height of almost 30,000 feet (10,000 meters) above SEA LEVEL.

Shire. English COUNTY.

Shoal. Underwater RIDGE or sandbed that reduces the water's depth to a point that might be unsafe for vessels.

Shock city. CITY that typifies disturbing changes in social and cultural conditions or in economic conditions. In the nineteenth century, the shock city of the United States was Chicago.

Shore. The zone where land and SEA meet. It extends from the water's edge at the lowest TIDE to the farthest point inland where SAND has been deposited by the largest STORM WAVES.

Shoreline. The specific PLACE where the land meets the SEA. Since WAVES and TIDES change constantly, the shoreline is not a fixed LOCATION. The position of MEAN SEA LEVEL is based on calculations of records of shoreline height.

Sial. Acronym for *si*lica and *al*umina. Those are the two principal constituents of light and crystalline ROCKS, such as GRANITE, that make up the greater part of the earth's CONTINENTAL CRUST. Heavier, basaltic rocks are referred to as SIMA.

Sierra. Spanish word for a mountain RANGE with a serrated crest. In California, the Sierra Nevada is an important range, containing Mount Whitney, the highest PEAK in the continental United States.

Silica. Oxide of silicon, with the chemical formula SiO_2. Silica occurs as QUARTZ or as part of many other ROCKS, including GRANITE. Silica is the most abundant oxide on Earth, and quartz is the second-most abundant MINERAL, after FELDSPAR.

Sill. Feature formed by INTRUSIVE volcanic activity. When LAVA is forced between two layers of ROCK, it can form a narrow horizontal layer of BASALT, parallel with the adjacent beds. Although it resembles a windowsill in its flatness, a sill may be hundreds of miles long and can range in thickness from a few centimeters to considerable thickness.

Silt. Intermediate TEXTURE size, for SOIL particles or for SEDIMENT, between SAND and CLAY. Silt particles have a diameter of 0.00016 to 0.0024 inch (0.004-0.06 millimeter). Silt is carried in SUSPENSION by RIVERS, giving them an opaque appearance, with color ranging from reddish to yellowish to brown-gray, depending on the MINERAL content of the silt. Silt can also be blown by WINDS. When it is deposited in thick layers, it is called LOESS.

Siltation. Build-up of SILT and SAND in creeks and waterways as a result of SOIL EROSION, clogging water courses and creating DELTAS at RIVER MOUTHS. Siltation often results from DEFORESTATION or removal of tree cover. Such ENVIRONMENTAL DEGRADATION causes loss of agricultural productivity, worsening of water supply, and other problems.

Sima. Acronym for *si*lica and *ma*gnesium. These are the two principal constituents of heavy ROCKS such as BASALT, which forms much of the OCEAN floor. Lighter, more abundant rock is SIAL.

Simple crater. Small IMPACT CRATER with a simple bowl shape.

Sinkhole. Circular depression in the ground surface, caused by WEATHERING of LIMESTONE, mainly through the effects of SOLUTION on JOINTS in the ROCK. If a STREAM flows above ground and then disappears down a sinkhole, the feature is called a swallow hole. In everyday language, many events that cause the surface to collapse are called sinkholes, even though they are rarely in limestone and rarely caused by weathering.

This sinkhole, which appeared in central Alabama in 1972, was 350 feet (105 meters) wide and 150 feet (45 meters) deep. (U.S. Geological Survey)

Sinking stream. STREAM or RIVER that loses part or all of its water to pathways dissolved underground in the BEDROCK.

Site. Locational attributes of a TOWN or CITY, its physical setting as well as its layout. In earlier times, a site was often chosen for its defensive property, so hilltops, or ISLANDS in RIVERS, became the sites of SETTLEMENTS.

Situation. Relationship between a PLACE, such as a TOWN or CITY, and its RELATIVE LOCATION within a REGION. A situation on the COAST is desirable in terms of overseas TRADE.

Slate is a metamorphic rock that can be split into thin sheets. (PhotoDisc)

Slate. METAMORPHIC ROCK that has a unique ability to be split into thin sheets; some slates are resistant to WEATHERING and are thus good for exterior use.

Sleet. Transparent drops of ice, caused by the freezing of raindrops. A TEMPERATURE INVERSION with below-freezing temperatures near the earth surface is a common cause of sleet. Sometimes, a mixture of SNOW and rain is incorrectly referred to as sleet.

Slip-face. LEEWARD side of a SAND DUNE. As the WIND piles up sand on the WINDWARD side, it then slips down the rear or slip-face. The angle of the slip-face is gentler than the angle of the windward slope.

Slough. Depression of the earth's surface containing a small amount of water and mud; a kind of MARSH or BOG.

Slump. Type of LANDSLIDE in which the material moves downslope with a rotational motion, along a curved slip surface.

Smog. Composite word formed from *sm*oke and f*og*. It was originally coined to describe the foul combination in London in the nineteenth century, when COAL fires were heavily used to heat homes and power factories. Sulfur dioxide, produced by burning coal, emitted sulfuric acid into the moist ATMOSPHERE. This true fog is also called industrial smog, to distinguish it from PHOTOCHEMICAL SMOG, which is a misnomer, because it involves no smoke.

Snout. Terminal end of a GLACIER.

Snow is frozen water in a crystalline form. More than one-fifth of the earth's land surface is covered in snow or ice. (PhotoDisc)

Snow. Frozen water in a crystalline form. Snowflakes have a hexagonal shape and form at high ALTITUDES around tiny nuclei such as DUST particles. More than one-fifth of the earth's land surface is covered in snow or ice.

Snow line. The height or ELEVATION at which snow remains throughout the year, without melting away. Near the EQUATOR, the snow line is more than 15,000 feet (almost 5,000 meters); at higher LATITUDES, the snow line is correspondingly lower, reaching SEA LEVEL at the POLES. The actual snow line varies with the time of year, retreating in summer and coming lower in winter.

Social Darwinism. Application of the ideas of Charles Darwin to human societies. Darwin thought that animal organisms or species evolved as the result of the struggle to survive in their physical ENVIRONMENT; social Darwinists believe that human groups also struggle to survive in particular environments. Cultural groups thus evolved through their ability to adjust and adapt to their physical environment. These ideas are closely related to the theory of ENVIRONMENTAL DETERMINISM.

Soil. The fine, natural material covering most of the earth's land surface, in which plants grow. Soil is formed by the physical and CHEMICAL WEATHERING of ROCK. Organisms ranging from bacteria and algae to worms, insects, and rodents make their home in soil. Soil is a mixture of MINERALS and organic matter, containing both water and AIR. As the basis of plant life, soil supports all terrestrial life on Earth. Soils take thousands of years to form but can be degraded or eroded rapidly, so soil conservation is a major area of concern throughout the world.

Soil horizon. SOIL consists of a series of layers called horizons. The uppermost layer, the O horizon, contains organic materials such as decayed leaves that have been changed into HUMUS. Beneath this is the A horizon, the TOPSOIL, where farmers plow and plant seeds. The B HORIZON often contains MINERALS that have been washed downwards from the A horizon, such as calcium, iron, and aluminum. The A and B horizons together comprise a solum, or true soil. The C horizon is weathered BEDROCK, which contains pieces of the original ROCK from which the soil formed. Another name for the C horizon is REGOLITH. Beneath this is the R horizon, or bedrock.

Soil moisture. Water contained in the unsaturated zone above the WATER TABLE.

Soil profile. Vertical section of a SOIL, extending through its horizon into the unweathered parent material.

Soil stabilization. Engineering measures designed to minimize the opportunity and/or ability of EXPANSIVE SOILS to shrink and swell.

Solar constant. Average value for the INSOLATION received at the THERMOPAUSE, or outer limit of the earth's ATMOSPHERE. The solar constant is 1,372 watts per square meter (2 calories per square centimeter per minute, or 2 langleys per minute).

Solar eclipse. See ECLIPSE, SOLAR.

Solar energy. One of the forms of ALTERNATIVE or RENEWABLE ENERGY. In the late 1990's, the world's largest solar power generating plant was located at Kramer Junction, California. There, solar energy heats huge OIL-filled containers with a parabolic shape, which produces steam to drive generating turbines. An alternative is the production of energy through photovoltaic cells, a TECHNOLOGY that was first developed for space exploration. Many individual homes, especially in isolated areas, use this technology.

Solar nebula. Disk-shaped cloud of hot DUST and gas from which the SOLAR SYSTEM formed.

Solar radiation. Transfer of heat from the SUN to the earth's surface, where it is absorbed and stored. See also INSOLATION.

Solar system. SUN and all the bodies that ORBIT it, including the PLANETS and their SATELLITES, plus numerous COMETS, ASTEROIDS, and METEOROIDS.

Solar wind. Gases from the SUN'S ATMOSPHERE, expanding at high speeds as streams of charged particles.

Solifluction. Word meaning flowing SOIL. In some REGIONS of PERMAFROST, where the ground is permanently frozen, the uppermost layer thaws during the summer, creating a saturated layer of soil and REGOLITH above the hard layer of frozen ground. On slopes, the material

Solifluction lobes on the side of a kame in the Nunatarssuaq region of Greenland. (U.S. Geological Survey)

can flow slowly downhill, creating a wavy appearance along the hillslope.

Solstices. Dates on which the Sun's rays at noon are vertically above the tropics, which are at their SUBSOLAR POINTS. The WINTER SOLSTICE in the NORTHERN HEMISPHERE occurs on December 21 or 22; this is the shortest DAY of the year in that HEMISPHERE. The summer solstice in the Northern Hemisphere occurs on June 20 or 21. The subsolar point then is the tropic of CANCER, and this is the longest day of the year for the Northern Hemisphere. In the SOUTHERN HEMISPHERE, the solstices occur on the same day, but the SEASONS are reversed: winter begins on the June solstice and summer begins on the December solstice.

Solution. Form of CHEMICAL WEATHERING in which MINERALS in a ROCK are dissolved in water. Most substances are soluble, but the combination of water with CARBON DIOXIDE from the ATMOSPHERE means that RAINFALL is slightly acidic, so that the chemical reaction is often a combination of solution and carbonation.

Sound. Long expanse of the SEA, close to the COAST, such as a large ESTUARY. It can also be the expanse of sea between the mainland and an ISLAND.

Source rock. ROCK unit or bed that contains sufficient organic carbon and has the proper thermal history to generate OIL or gas.

Southern Hemisphere. The half of the earth below the EQUATOR.

Southern Oscillation. Atmospheric "seesaw" that tilts between ATMOSPHERIC PRESSURE extremes at Tahiti and Darwin, Australia.

Sovereignty. Exercise of government and STATE power over people and the territory they occupy. Sovereignty is recognized by other sovereign states and is upheld by international law. The individual states of the United States, the PROVINCES of Canada, and the counties of the United Kingdom are administrative subdivisions of these independent countries. These smaller entities do not have sovereignty.

Spa. PLACE with natural MINERAL SPRINGS.

Spatial diffusion. Notion that things spread through space and over time. An understanding of geographic change depends on this concept. Spatial diffusion can occur in various ways. Geographers distinguish between expansion diffusion, relocation diffusion, and hierarchical diffusion.

Spheroidal weathering. Form of ROCK WEATHERING in which layers of rock break off parallel to the surface, producing a rounded shape. It results from a combination of physical and CHEMICAL WEATHERING. Spheroidal weathering is especially common in GRANITE, leading to the creation of TORS and similar rounded features. Onion-skin weathering is a term sometimes used, especially when this is seen on small rocks.

Spillway. Generally, a broad reinforced CHANNEL near the top of the DAM, designed to allow rising waters to escape the RESERVOIR without over-topping the dam.

Spit. Long, narrow sandbar extending outward from the COAST. A sand spit is attached to the coast at one end. Cape Cod is a famous spit. See also HOOK.

Barrier sand spit along the South Carolina coast. (U.S. Geological Survey)

Splash erosion. EROSION that occurs when raindrops hit the ground, dislodging particles of SOIL or weathered material and causing them to move downslope. Splash erosion can lead to OVERLAND FLOW, which can cause considerable erosion of newly plowed ground.

Spread effects. Positive impacts on economic growth throughout a REGION. Economic growth in a center or region is usually accompanied by spread effects. For example, the effect of providing work and income leads to an increased demand for housing, food, entertainment, and other consumer goods, thereby creating further employment and growth.

Spring. PLACE where water flows naturally from the ground, found wherever the WATER TABLE intersects the earth's surface; in KARST TOPOGRAPHY, a spring represents the discharge point of a CAVE.

Spring sapping. Process in which water flows out of subsurface SPRINGS to surface level, forming a STREAMBED as it flows downslope.

Spring tide. TIDE of maximum RANGE, occurring when lunar and solar tides reinforce each other, a few DAYS after the full and new MOONS.

Squall line. Line of vigorous THUNDERSTORMS created by a cold downdraft that spreads out ahead of a fast-moving COLD FRONT.

Squatter settlements. URBAN residential slums built by recent urban immigrants on land that they do not own or rent. Shacks in the squatter settlements are built of found materials, including cardboard, mud, grass, and plastic sheeting. These squatter settlements are known by different names in different countries: "favelas" in Brazil, "callampas" in Chile, "villas miserias" in Argentina, "bustees" in India, and "gourbevilles" or "bidonvilles" in parts of Africa. Governments often supply water and power to squatter settlements, and residents may form communities to improve the structures and services. Also called informal settlements.

Stacks. Pieces of ROCK surrounded by SEA water, which were once part of the mainland. WAVE EROSION has caused them to be isolated. Also called sea stacks.

Sea stacks. (Corbis)

Stalactites (above) and stalagmites (below) that have grown together in Kentucky's Mammoth Cave National Park. (U.S. Geological Survey)

Stalactite. Long, tapering piece of calcium carbonate hanging from the roof of a LIMESTONE CAVE or cavern. Stalactites are formed as water containing the MINERAL in solution drips downward. The water evaporates, depositing the dissolved minerals. See also STALAGMITE.

Stalagmite. Column of calcium carbonate growing upward from the floor of a LIMESTONE CAVE or cavern. See also STALACTITE.

State. Territory and its political organization, with administration regulated by a government with sovereign powers, and that is recognized as legitimate by other states that are members of the international community of legitimate states. The international BORDERS of the state must be agreed upon by adjacent states and by other states. A state has a citizen POPULATION resident within its territory and an organized and functioning ECONOMY. In the United States, "state" is also the term used for a subdivision of the whole; in other countries, such an internal administrative REGION is called a PROVINCE, department, or other name.

Steppe. Huge REGION of GRASSLANDS in the midlatitudes of Eurasia, extending from central Europe to northeast China. The region is not uniform in ELEVATION; most of it is rolling PLAINS, but some mountain RANGES also occur. These have not been a barrier to the migratory lifestyle of the herders who have occupied the steppe for many centuries. The Asian steppe is colder than the European steppe, because of greater elevation and greater continentality. The best-known rulers from the steppe were the Mongols, whose empire flourished in the thirteenth and fourteenth centuries. Geographers speak of a steppe

CLIMATE, a semiarid climate where the EVAPORATION rate is double that of PRECIPITATION. South of the steppe are great DESERTS; to the north are midlatitude mixed FORESTS. In terms of climate and VEGETATION, the steppe is like the short-grass PRAIRIE vegetation west of the Mississippi River. Also called steppes.

Stock. Feature formed by INTRUSIVE volcanic activity. LAVA rises toward the surface and forms a mass or POOL that slowly cools into a granitic ROCK LANDFORM, often circular in shape. Removal of the overlying materials can subsequently expose the stock as a HILL. A much larger feature formed in the same manner is a BATHOLITH.

Stone river. See ROCK GLACIER.

Storm. Atmospheric disturbance with rotating WINDS of considerable speed, associated with lower-than-usual pressure. CLOUDS, PRECIPITATION, and often thunder and LIGHTNING accompany the passage of a storm. Storms can be classified as HURRICANES, TORNADOES, or low-pressure systems.

Storm clouds forming over the Grand Tetons in Wyoming. (PhotoDisc)

Storm surge. General rise above normal water level, resulting from a HURRICANE or other severe coastal STORM.

Strait. Relatively narrow body of water, part of an OCEAN or SEA, separating two pieces of land. The world's busiest SEAWAY is the Johore Strait between the Malay Peninsula and the island of Sumatra.

Strata. Layers of SEDIMENT deposited at different times, and therefore of different composition and TEXTURE. When the sediments are laid down, strata are horizontal, but subsequent tectonic processes can

Folds in strata at the southern end of Montana's Scapegoat Mountain. (U.S. Geological Survey)

lead to tilting, FOLDING, or faulting. Not all SEDIMENTARY ROCKS are stratified. Singular form of the word is stratum.

Strategic resources. RESOURCES considered essential for a NATION's major industries, military defense, and ENERGY programs. For the United States, these resources include manganese, chromium, cobalt, nickel, platinum, titanium, aluminum, and OIL.

Stratified drift. Material deposited by glacial MELTWATERS; the water separates the material according to size, creating layers.

Stratigraphic time scale. History of the evolution of life on the earth broken down into time periods based on changes in FOSSIL life in the sequence of ROCK layers; the time periods were named for the localities in which they were studied or from their characteristics.

Stratigraphic unit. Any ROCK layer that can be easily recognized because of specific characteristics, such as color, composition, or grain size.

Stratigraphy. Study of sedimentary STRATA, which includes the concept of time, possible correlation of the ROCK units, and characteristics of the rocks themselves.

Stratosphere. Layer of the ATMOSPHERE distinguished by a rise in TEMPERATURE from bottom to top. This warming mainly results from absorption of SOLAR RADIATION by OZONE molecules found in the stratosphere. The stratosphere extends from about 11 miles (17 km.) above the earth's surface to about 30 miles (50 km.) in ALTITUDE. Below the stratosphere is the TROPOSPHERE; above it is the MESOSPHERE.

Mount Rainier, an ancient stratovolcano, one whose eruptions are of different types and produce different lavas. (Corbis)

Stratovolcano. Type of VOLCANO in which the ERUPTIONS are of different types and produce different LAVAS. Sometimes an eruption ejects cinder and ASH; at other times, viscous lava flows down the sides. The materials flow, settle, and fall to produce a beautiful symmetrical LANDFORM with a broad circular base and concave slopes tapering upward to a small circular CRATER. Mount Rainier, Mount Saint Helens, and Mount Fuji are stratovolcanoes. Also known as a COMPOSITE CONE.

Stratum. A single bed or layer of SEDIMENTARY ROCK. See also STRATA.

Stream. Body of water in a CHANNEL, moving downhill because of GRAVITY. Geographers prefer the term to "river," because it emphasizes the fact

Geographers and other physical scientists tend to prefer the word "stream" over "river" for a body of water that moves down a channel. (PhotoDisc)

that water flows in a confined channel, between BANKS. "River" is a less precise term, partly because it suggests a large and constant stream. In arid REGIONS, streams are often INTERMITTENT in their flow, or even EPHEMERAL, when they contain flowing water only for a short period.

Stream order. System of studying STREAMS devised by Robert Horton, an American hydrologist. The smallest streams in the HEADWATERS are designated first-order streams. When two first-order streams converge, the result is a second-order stream. When two second-order streams converge, a third-order stream is formed. Quantifying a DRAINAGE network in this way enables calculations of drainage area, discharge, and other factors in the stream network.

Streambed. Channel through which a STREAM flows. Dry streambeds are variously known as ARROYOS, DONGAS, WASHES, and WADIS.

Striations cut into Devonian-era dolomite by large, sharp rocks pushed by glacial ice. (U.S. Geological Survey)

Striations. Grooves eroded into BEDROCK by the ground MORAINE, or ROCKS, carried by a GLACIER as it makes its way downslope. Sometimes the striations are merely scratches; in other places the grooves can be several centimeters deep. Study of striations now exposed reveals much about the direction and size of glacial flows during previous ICE AGES.

Strike. Term used when earth scientists study tilted or inclined beds of SEDIMENTARY ROCK. The strike of the inclined bed is the direction of a horizontal line along a bedding plane. The strike is at right angles to the dip of the rocks.

Strike-slip fault. In a strike-slip fault, the surface on either side of the fault moves in a horizontal plane. There is no vertical displacement to form a FAULT SCARP, as there is with other types of faults. The San Andreas Fault is a strike-slip fault. Also called a transcurrent fault.

Strip mining. Removal of a long narrow strip of surface materials, using excavation machinery called a dragline. The underlying MINERAL deposit then can be collected easily. When the dragline moves across to the adjacent land to excavate the next strip, parallel to the first, the waste or overburden from the former strip is deposited back over that strip of land. COAL deposits are often mined using strip mining. This type of mining is destructive of natural ENVIRONMENTS.

Strip mining involves the removal of long narrow strips of surface materials, using excavation equipment called draglines, so that underlying minerals can be collected easily. (PhotoDisc)

Subcontinent. Large piece of a CONTINENT. The term is especially used when referring to the Indian subcontinent.

Subduction. Process that occurs when two TECTONIC PLATES converge. If one plate is composed of lighter CONTINENTAL CRUST and the other of heavier OCEANIC CRUST, the lighter plate rides up over the heavier plate, forcing it downward. This is a destructive process, destroying crust. At PLATE BOUNDARIES where subduction is occurring, oceanic TRENCHES are found close to the SHORE, with tall, young MOUNTAINS close to the COAST on the land. Active VOLCANOES are common, and ERUPTIONS and EARTHQUAKES are frequent. The combination of the Peru Trench and the Andes Mountains marks a large plate boundary REGION where subduction is proceeding.

Subduction zone. CONVERGENT PLATE BOUNDARY where an oceanic PLATE is being thrust below another plate.

Sublimation. Process by which water changes directly from solid (ice) to vapor, or vapor to solid, without passing through a liquid stage.

Submarine canyon. CHANNEL cut deep in the seafloor SEDIMENTS by RIVERS or submarine CURRENTS.

Submergence. COASTLINE of submergence is formed when SEA LEVELS that have risen since the last ICE AGE have made former RIVER VALLEYS and other LANDFORMS INLETS, estuaries, and BAYS.

Subsidence. Sinking of the earth's surface or a decrease in the distance between the earth's surface and its center.

Subsistence agriculture. System of production in which farmers grow only enough food to feed themselves and their immediate families, with just enough seed left over to ensure a crop the following year. No surplus is produced for sale. Shifting cultivation is a form of subsistence agriculture. Subsistence agriculturalists usually produce a variety of crops throughout the year and may keep a few animals for food. The opposite of subsistence agriculture is commercial agriculture.

Subsoil. Term for the C horizon in a soil. See also SOIL.

Subsolar point. Point on the earth's surface where the SUN is directly overhead, making the Sun's rays perpendicular to the surface. The subsolar point receives maximum INSOLATION, compared with other PLACES, where the Sun's rays are oblique.

Suburbanization. Growth of POPULATION around the edge or fringe of a CITY. This process of city growth began in the United States in the eighteenth century, when wealthy people moved out of the crowded, unhealthful city to the RURAL edge where they could have a large property with fresh AIR and a large garden. This led to recent immigrants moving into the inner city.

Sultanate. Islamic STATE ruled by a person with the title of sultan.

Summer solstice. See SOLSTICES.

Summit. Highest part of any LANDFORM remnant, HILL, PEAK, or MOUNTAIN. A summit can be either smooth or sharply defined.

Sun. The center of Earth's SOLAR SYSTEM, the Sun is an average star in terms of its physical characteristics. It is a large sphere of incandescent gas that has a diameter more than 100 times that of Earth, a mass more than 300,000 times that of Earth, and a volume 1.3 million times that of Earth. The Sun's surface GRAVITY is thirty-four times that of Earth. The earth revolves around the Sun in a slightly elliptical ORBIT that takes exactly one Earth year to complete.

Sun Belt. Name given to certain parts of the United States that attract IMMIGRATION because of their warm sunny CLIMATES. Retired people are the main component of Sun Belt MIGRATION. The Sun Belt is not a continuous BELT, but California, Texas, Florida, Nevada, and New Mexico have benefited from this trend.

Sunrise occurs when the top of the Sun first appears above the horizon. As with sunsets, these times vary with location and season. (PhotoDisc)

Sunrise. The time when the top of the SUN first appears above the HORIZON. This time changes throughout the year.

Sunset. The time when the last part of the SUN totally disappears below the HORIZON. This time changes throughout the year.

Sunset is the moment when the last part of the Sun disappears below the horizon. The exact times vary within individual time zones and change throughout the year, with the changes becoming greater with distance away from the equator. (PhotoDisc)

Sunspots. REGIONS of intense magnetic disturbances that appear as dark spots on the solar surface; they occur approximately every eleven years.

Supercontinent. Vast LANDMASS of the remote geologic past formed by the collision and amalgamation of crustal PLATES. Hypothesized supercontinents include PANGAEA, GONDWANALAND, and LAURASIA.

Supersaturation. State in which the AIR'S RELATIVE HUMIDITY exceeds 100 percent, the condition necessary for vapor to begin transformation to a liquid state.

Supranationalism. Process by which autonomous countries join together in an agreement for their mutual benefit. Supranational ventures usually are economic in nature; the North American Free Trade Agreement is an example of supranationalism between the United States, Canada, and Mexico. Supranationalism can also be cultural or military, for example the North Atlantic Treaty Organization.

Supratidal. Referring to the SHORE area marginal to shallow OCEANS that are just above high-tide level.

Surface water. Relatively warm seawater between the OCEAN surface and that depth marked by a rapid reduction in TEMPERATURE.

Suspension. Means by which small particles are moved by water or WIND. The particles are so light that they can be picked up and transported. Suspended particles make RIVERS appear muddy, or make DUST clouds visible.

Sustainable agriculture. Commercial agriculture that is ecologically responsible and sound. SOIL conservation is practiced, using contouring and shallow plowing. Pest management is achieved by INTERCROPPING or natural pesticides, rather than chemicals.

Sustainable energy. See ALTERNATIVE ENERGY.

Swamp. WETLAND where trees grow in wet to waterlogged conditions. Swamps are common close to the RIVER on FLOODPLAINS, as well as in some coastal areas.

Swell. Regular pattern of smooth rounded WAVES moving across the OCEAN surface in one direction.

Swidden. Area of land that has been cleared for SUBSISTENCE AGRICULTURE by a farmer using the technique of slash-and-burn. A variety of crops is planted, partly to reduce the risk of crop failure. Yields are low from a swidden because SOIL fertility is low and only human labor is used for CLEARING, planting, and harvesting. See also INTERTILLAGE.

Symbiosis. Cooperative living arrangement of two different species. When both species benefit, it is called mutualism; when one benefits more than the other, it is called commensalism. Parasitism sometimes is regarded as a form of symbiosis.

Symbolic landscapes. LANDSCAPES centered on buildings or structures that are so visually emblematic that they represent an entire CITY. The Eiffel Tower of Paris or the Harbour Bridge of Sydney are examples of such features. Other cities have more generic cityscapes that are symbolic of the entire NATION or the entire CULTURE. For the United States, three such symbolic cityscapes are recognized by most geographers. The New England VILLAGE or townscape, with the steepled white wooden church and village green, accompanied by DECIDUOUS trees in fall colors, not only represents a regional architecture but also symbolizes a community rooted in Puritan values of morality, industriousness, and a God-centered, family-oriented life. The familiar symbolic cityscape generally referred to as Main Street U.S.A. is an image of an earlier age, with sidewalks and small, family-run shops, which was adopted as the centerpiece of Disneyland. The California landscape is a third symbolic U.S. landscape, typified by tall palms, suburban houses, and an individualistic, recreation-oriented, middle-class POPULATION. This final landscape has been widely popularized by the motion picture industry.

Syncline. Downfold or TROUGH shape that is formed through compression of ROCKS. An upfold is an ANTICLINE.

Syncline in weathered shale, with a left limb that dips to the right and a vertical right limb. (U.S. Geological Survey)

Cape Town's aptly named Table Mountain, viewed from Table Bay in South Africa. (Corbis)

Table mountain. MESA with a particularly well-defined shape resembling a table. The most famous MOUNTAIN of this type is Cape Town's aptly named Table Mountain in South Africa.

Tableland. Large area of land with a mostly flat surface, surrounded by steeply sloping sides, or ESCARPMENTS. A small PLATEAU.

Taiga. Russian name for the vast BOREAL FORESTS that cover Siberia. The marshy ground supports a tree VEGETATION in which the trees are CO-NIFEROUS, comprising mostly pine, fir, and larch.

Takeoff. Stage in the economic DEVELOPMENT of a COUNTRY when conditions are right for the country to undergo an industrial revolution, making that country an industrialized export ECONOMY. The term comes from the work of the American economist Walter Rostow.

Talus. Broken and jagged pieces of ROCK, produced by WEATHERING of steep slopes, that fall to the base of the slope and accumulate as a talus cone. In high MOUNTAINS, a ROCK GLACIER may form in the talus. See also SCREE.

Tarn. Small circular LAKE, formed in a CIRQUE, which was previously occupied by a GLACIER.

Taxonomy. Another name for a system of scientific classification. The SOIL classification used in the United States is called the Soil Taxonomy or the Seventh Approximation.

Technology. Practical application of knowledge. It could refer to simple techniques such as using fire to cook food or a ROCK to crack oysters. In modern use, technology implies the use of power and machinery, as in mining technology or COMMUNICATIONS technology. In developed or high-income economies, technology has largely replaced the need for human labor, but many low-income countries have a low level of technology.

Tectonic plate. Large portion of the earth's CRUST. Plates are in constant motion, separating or colliding, changing the shape of CONTINENTS and the configuration of the surface both above and below SEA LEVEL. The North American Plate is slowly moving northwest. The other large plates are the South American, African, Eurasian, Indo-Australian, and ANTARCTIC plates. There are also several smaller plates.

Tectonism. The formation of MOUNTAINS because of the deformation of the CRUST of the earth on a large scale.

Temperate zone. Areas between the tropic of CANCER and the Arctic Circle and between the tropic of CAPRICORN and the Antarctic Circle. The ancient Greeks divided the world into three CLIMATE zones, based on their understanding of geometry and geography. At the EQUATOR was the hot REGION the Greeks called the Torrid Zone, believing that human life was not possible there. At the POLES were the two FRIGID ZONES, thought to be too cold for human life. Between the Torrid and Frigid Zones lay two temperate zones, one in each HEMISPHERE. The civilized world as known to the ancient Greeks lay in the temperate zone, where humans flourished because AGRICULTURE was possible. Although the Greeks believed that life was possible in the temperate zone of the SOUTHERN HEMISPHERE, they did not think that the beings there would resemble the humans of the NORTHERN HEMISPHERE. Because the Greeks thought it was not possible to travel through the Torrid Zone to see what the southern regions were like, there were speculations about whether the inhabitants there had one eye, or perhaps walked upside down, because of the earth's curvature. Drawings of creatures with feet on their heads led to the origin of the word ANTIPODES, which is now used to described southern lands.

Temperature. Measure of the kinetic ENERGY of molecules, felt by humans as sensible heat. Temperature is usually measured using the CELSIUS SCALE, but in the United States, the FAHRENHEIT SCALE is more commonly used. The Kelvin scale is used by scientists.

Temperature inversion. Increase in AIR TEMPERATURE with increased ALTITUDE. This is the opposite of normal conditions whereby the temperature in the TROPOSPHERE decreases uniformly with height. An inversion can be produced in a number of ways. A RADIATION or ground inversion occurs on cold clear nights when the ground cools rapidly through terrestrial radiation. Air in contact with this cold surface is then cooled, becoming colder than the air above it. If the air is moist and the temperature falls below the DEW POINT, FOG can form; this is called a radiation fog. A SUBSIDENCE inversion forms when air in a high-pressure cell descends and unequal compression causes the upper part of the air to become warmer than the lower part. An inversion can form in the surface layer or at an upper level in the troposphere. A layer of stratus CLOUDS usually marks the upper-level inversion. TO-POGRAPHY can be an important influence on a ground inversion. In hilly or mountainous areas, cold air drains into the VALLEYS, especially at night, and causes an inversion that can persist for more than one DAY. Under normal conditions, warm air near the surface rises because it is less dense, which lessens POLLUTION. Cool surface air overlain with warmer air prevents the upward rise of smoke and other pollutants, which become trapped under the inversion layer. The United States' worst AIR POLLUTION disaster occurred in Donora, Pennsylvania, in 1948, when a temperature inversion led to a deadly fog full of industrial pollutants; it persisted for four days, affecting thousands of people and causing twenty deaths.

Temporary base level. STREAMS or RIVERS erode their beds down toward a BASE LEVEL—in most cases, SEA LEVEL. A section of hard ROCK may slow EROSION and act as a temporary, or local, base level. Erosion slows upstream of the temporary base level. A DAM is an artificially constructed temporary base level.

Tension. Type of stress that produces a stretching and thinning or pulling apart of the earth's CRUST. If the surface breaks, a NORMAL FAULT is created, with one side of the surface higher than the other.

Tephra. General term for volcanic materials that are ejected from a vent during an ERUPTION and transported through the AIR, including ASH (volcanic), BLOCKS (volcanic), cinders, LAPILLI, scoria, and PUMICE.

Terminal moraine. RIDGE of unsorted debris deposited by a GLACIER. When a glacier erodes it moves downslope, carrying ROCK debris and

Terminal moraine in the French Alps. (Mark Twain, *A Tramp Abroad*, 1880)

creating a ground MORAINE of material of various sizes, ranging from big angular blocks or boulders down to fine CLAY. At the terminus of the glacier, where the ice is melting, the ground moraine is deposited, building the ridge of unsorted debris called a terminal moraine.

Terra rossa. Red SOIL formed from LIMESTONE, which provides a strong contrast to the paler limestone BEDROCK below. The red color comes from insoluble iron hydroxides. Name is Italian for red soil.

Terrace. Horizontal RIDGE in a hillside. In many Asian countries, the steep slopes of HILLS or mountainsides have been transformed, through great human effort, into a series of steplike terraces to provide flat land for rice PADDIES. RIVER TERRACES are natural formations on either side of a RIVER.

Asian rice paddies maximizing agriculture use of difficult mountainous terrain. (PhotoDisc)

Terracettes. Small parallel TERRACES or steps on a hillslope. They are thought to originate from a combination of mass movement downslope and trampling by the hooves of grazing animals.

Terrain. Physical features of a REGION, as in a description of rugged terrain. It should not be confused with TERRANE.

Terrane. Piece of CONTINENTAL CRUST that has broken off from one PLATE and subsequently been joined to a different plate. The terrane has quite different composition and structure from the adjacent continental materials. Alaska is composed mostly of terranes that have accreted, or joined, the North American plate.

Terrestrial planet. Any of the solid, rocky-surfaced bodies of the inner SOLAR SYSTEM, including the PLANETs Mercury, Venus, Earth, and Mars and Earth's SATELLITE, the MOON.

Terrigenous. Originating from the WEATHERING and EROSION of MOUNTAINS and other land formations.

Tertiary industry. Sector of the ECONOMY, also known as the service sector, that does not produce material goods for sale. Tertiary industry includes services such as banking and insurance, real estate, retailing, TRANSPORTATION and COMMUNICATIONS, and such necessities as police, defense, and education. In high-income economies, most people are employed in the tertiary sector; the secondary sector is of diminished importance, and primary industries are highly mechanized, with few workers.

Tertiary period. PERIOD in the CENOZOIC ERA of the geologic time scale; it encompasses the time span between about 65 million and 2 million years ago.

Texture. One of the properties of SOILS. The three textures are SAND, SILT, and CLAY. Texture is measured by shaking the dried soil through a series of sieves with mesh of reducing diameters. A mixture of sand, silt, and clay gives a LOAM soil.

Thalweg. Profile obtained when the ELEVATION of a STREAMBED is plotted against the STREAM's length. The shape is a concave curve. German word for "valley way" or "path," spelled *Talweg* in modern German.

Thematic map. MAP that displays information concerning a theme, such as geology, VEGETATION, or annual PRECIPITATION. Using computer mapping with a Geographic Information System (GIS), many themes can be added to a project, overlaid to create spatial queries, and displayed as desired.

Thermal equator. Imaginary line connecting all PLACES on Earth with the highest mean daily TEMPERATURE. The thermal equator moves south of the EQUATOR in the SOUTHERN HEMISPHERE summer, especially over the CONTINENTS of South America, Africa, and Australia. In the northern summer, the thermal equator moves far into Asia, northern Africa, and North America.

Thermal erosion. EROSION of water ice from a solid state to vapor.

Thermal fracture. Formation of a fracture or crack in a ROCK as a result of TEMPERATURE changes.

Thermal gradient. Increase of TEMPERATURE with depth below the earth's surface, expressed as DEGREES Celsius per kilometer; the average is 25 to 30 degrees Celsius per kilometer; also known as GEOTHERMAL gradient.

Thermal pollution. Disruption of the ECOSYSTEM caused when hot water is discharged, usually as a thermal PLUME, into a relatively cooler body of water. The TEMPERATURE change affects the aquatic ecosystem, even if the water is chemically pure. Nuclear power-generating plants use large volumes of water in the process and are important sources of thermal pollution.

Thermal springs. See HOT SPRINGS.

Thermocline. Depth interval at which the TEMPERATURE of OCEAN water changes abruptly, separating warm SURFACE WATER from cold, deep water.

Thermodynamics. Area of science that deals with the transformation of ENERGY and the laws that govern these changes; equilibrium thermodynamics is especially concerned with the reversible conversion of heat into other forms of energy.

Thermometer. Instrument for measuring TEMPERATURE. Commonly, a long thin glass tube containing alcohol. Early thermometers used mercury in a glass tube. The liquid inside the tube expands when the temperature rises. There are three temperature SCALES in regular use: CELSIUS, FAHRENHEIT, and Kelvin. Most countries in the world use the Celsius scale, where the boiling point of pure water is 100 DEGREES and the freezing point is zero. Temperatures in the United States are usually given in degrees Fahrenheit. On this scale, the temperature at which water boils is 212 degrees, and the temperature at which water freezes is 32 degrees.

Thermopause. Outer limit of the earth's ATMOSPHERE.

Thermosphere. Atmospheric zone beyond the MESOSPHERE in which TEMPERATURE rises rapidly with increasing distance from the earth's surface.

Third World. Term formerly used to refer to low-income countries, where the standard of living was poor and per-capita income low. The term dates from the COLD WAR, when the capitalist countries were regarded as the First World, and the Communist countries as the Second World.

Threshold. Minimum market size required to make the sale of a product, or the provision of a service, economically profitable. Luxury goods may need few buyers and so they have a low threshold; fast-food outlets need to sell large quantities of their product, so they have a high threshold. See also RANGE.

Thrust belt. Linear BELT of ROCKS that have been deformed by THRUST FAULTS.

Thrust fault showing fault drag. (U.S. Geological Survey)

Thrust fault. FAULT formed when extreme compression of the earth's CRUST pushes the surface into folds so closely spaced that they overturn and the ROCK then fractures along a fault.

Thunderstorm. Huge CUMULONIMBUS CLOUD that brings heavy rain, or sometimes hail, together with thunder and LIGHTNING. CUMULUS clouds form in moist warm AIR as it rises, and the presence of updrafts can lead to continued growth of the clouds into a thunderhead. Such clouds typically have a flat top, or anvil head, when they reach their greatest height of development, which may be in the STRATOSPHERE. A TORNADO can develop from a thunderstorm. MICROBURSTS are another common phenomenon.

Cumulus clouds form in moist warm air as it rises, and the presence of updrafts can lead to continued growth of the clouds into thunderheads, which typically have flat tops, or anvil heads, when they reach their greatest elevations, which may be in the stratosphere. (Corbis)

Tidal bore. See BORE.

Tidal energy. The regular EBB and flow of TIDES can be harnessed, in suitable LOCATIONS such as narrow INLETS or estuaries where there is a large TIDAL RANGE, and used to generate electricity. The oldest tidal generating plant is located at La Rance in France; another is located in the Bay of Fundy in Nova Scotia.

Tidal force. Gravitational force whose strength and direction vary over a body and thus act to deform the body.

Tidal range. Difference in height between high TIDE and low tide at a given point.

Tidal wave. Common but inaccurate name for a TSUNAMI.

Tides. Daily variations in SEA LEVEL, and in large LAKES, caused by the gravitational pull of the MOON and SUN on the earth, and especially on the HYDROSPHERE. When Earth, Moon, and Sun are in conjunction (lined up), it causes SPRING TIDES with the greatest TIDAL RANGE (highest and lowest tides). When the three bodies are in opposition (aligned at right angles), it causes NEAP TIDES, those in which the tidal range is smallest. Most COASTS on Earth experience two high tides and two low tides in a 24-hour DAY.

High tide at Bolinas Lagoon in California's Marin County in 1906. (U.S. Geological Survey)

Low tide at Bolinas Lagoon in California's Marin County in 1906. (U.S. Geological Survey)

Till. Mass of unsorted and unstratified SEDIMENTS deposited by a GLA-CIER. Boulders and smaller rounded ROCKS are mixed with CLAY-sized materials.

Timberline. Another term for treeline, the BOUNDARY of tree growth on MOUNTAIN slopes. Above the timberline, TEMPERATURES are too cold for tree growth.

Time-space convergence. Concept explaining how TECHNOLOGY has enabled PLACES to seem closer, because the time to send a communication between two places, or the time to travel from one place to the other, has shortened. The history of transport is one of increased time-space convergence, as clipper ships, steam ships, trains, automobiles, and airplanes reduced the time of journeys. The radio, telegraph, telephone, and Internet allow for almost instantaneous communication, the ultimate in time-space convergence.

Time zones. The earth is divided into twenty-four standard time zones, each of which is fifteen DEGREES of LONGITUDE apart. The central MERIDIAN for the first time zone is the PRIME MERIDIAN, or zero degrees. Each central meridian is fifteen degrees apart, so all central meridians are a factor of five or ten degrees. Because political boundaries do not conform well to meridians in some PLACES, time zones do not follow meridians exactly, but are often adjusted to the political BOUNDARY.

Tombolo. Strip of SAND or other SEDIMENT that connects an ISLAND or

Mont-Saint-Michel, an ancient fortified island abbey, is connected to the mainland by a tombolo.
(PhotoDisc)

SEA stack to the mainland. Mont-Saint-Michel is linked to the French mainland by a tombolo.

Topocide. Death of a PLACE, usually the result of INDUSTRIALIZATION, mining, or URBANIZATION.

Topographic map. MAP showing the detailed shape of the land using contours, which are imaginary lines drawn at equal ELEVATION above SEA LEVEL, with a regular contour interval. On an American topographic map, for example, contours might be shown for 20, 40, and 60 feet, and so on, above mean sea level. A standard set of symbols and colors is used in the production of topographic maps, so that a LEGEND is not necessary once a user becomes familiar with these maps. Topographic maps are used by hikers, campers, and engineers. The U.S. Geologic Survey produces topographic maps at SCALES of 1:24,000 and smaller scales.

Topography. Description of the natural LANDSCAPE, including LAND-FORMS, RIVERS and other waters, and VEGETATION cover.

Topological space. Space defined in terms of the connectivity between LOCATIONS in that space. The nature and frequency of the connections are measured, while distance between locations is not considered an important factor. An example of topological space is a transport network diagram, such as a bus route or a MAP of an underground rail system. Networks are most concerned with flows, and therefore with connectivity.

Toponyms. PLACE names. Sometimes, names of features and SETTLE-MENTS reveal a good deal about the history of a REGION. For example, the many names starting with "San" or "Santa" in the Southwest of the United States recall the fact that Spain once controlled that area. The scientific study of place names is toponymics.

Topophilia. Love of PLACE. Feelings or emotions that people associate with certain places. The home area or REGION hold a special place in the affections of many people and give them a sense of identity and belonging to a community.

Topsoil. In reclamation, all SOIL which will support plant growth, but normally the 8 to 12 inches (20-30 centimeters) of the organically rich top layer.

Tor. Rocky outcrop of blocks of ROCK, or corestones, exposed and rounded by WEATHERING. Tors frequently form in GRANITE, where three series of JOINTS often developed as the rock originally cooled when it was formed.

Tornado. Narrow vortex of rotating WINDS around a low-pressure center. Tornadoes are about 600 feet (200 meters) in diameter at ground level, and they travel across the land at speeds of up to 30 miles (50 km.) per hour. Within the tornado, windspeeds on average reach 270 miles (450 km.) per hour but can be even higher. Because of this, tornadoes are extremely destructive. Tornadoes can occur in many coun-

Narrow vortexes of wind rotating around low-pressure centers, tornadoes move across land surfaces at speeds of up to 30 miles (50 km.) per hour. (PhotoDisc)

tries, but the United States has the world's greatest frequency of tornadoes, especially in its Great Plains STATES.

Town. URBAN SETTLEMENT with a form of local self-government, such as a mayor. A town usually has more than twenty-five hundred residents and can be much larger. See also CITY.

Township and range. System of surveying and subdividing land quickly, introduced in the United States in 1785. Each township was a square with a side of six miles, or thirty-six square miles. Each square mile was then divided into four squares, each covering 160 acres. This was the smallest piece of land a farmer could buy. The legacy of this survey system is the checkerboard LANDSCAPE of the agricultural land of the Midwest.

Traction. Means by which a STREAM moves part of its load. Large PEBBLES, or even boulders, are dragged along, in contact with the bed of the stream. The process is traction; the material is the bedload.

Trade. Exchange of goods and services, with or without the use of currency. In modern economies, currency or money is the medium of trade. Trade opened the world to European influences, as mariners and explorers sought new sources of MINERALS and other trade goods. International trade is an important part of high-income economies and is regulated by agreements such as the General Agreement on Tariffs and Trade (GATT) and the World Trade Organization (WTO). Arguments for increased international trade include the creation of jobs in other countries that supply goods to wealthy markets, and the lower price of commodities in those countries. Arguments against international trade include the perpetuation of low-wage labor in poor countries and ENVIRONMENTAL DEGRADATION in some industries.

Trade winds. WINDS that converge toward the INTERTROPICAL CONVERGENCE ZONE. Trade winds move from the subtropical high-pressure zones of each HEMISPHERE toward the low-pressure BELT but are deflected by the CORIOLIS EFFECT and by friction, so that they produce the northeast trade winds in the NORTHERN HEMISPHERE and the southeast trade winds in the SOUTHERN HEMISPHERE. The name comes from the days when sail-powered ships carried goods between CONTINENTS. These warm and reliable winds were favored by sailors. Part of the circulation pattern known as Hadley cells.

Transculturation. Cultural mingling that occurs when two CULTURES are in close contact over a sustained period. The culture of modern Mexico, which combines Spanish and Amerindian cultures, is a good example of transculturation. Compare with ACCULTURATION.

Transferability. Economic term that describes the ability to move goods from one PLACE to another and to bear the costs incurred.

Transform faults. FAULTS that occur along DIVERGENT PLATE boundaries, or SEAFLOOR SPREADING zones. The faults run perpendicular to the spreading center, sometimes for hundreds of miles, some for more than five hundred miles. The motion along a transform fault is lateral or STRIKE-SLIP.

Transgression. Flooding of a large land area by the SEA, either by a regional downwarping of continental surface or by a global rise in SEA LEVEL.

Transhumance. Form of pastoral activity in which farmers take their grazing animals up to high ALPINE pastures during the spring, bringing them down to lower levels in the colder months. In cold CLIMATES, LIVESTOCK can even be kept indoors during the winter. Transhumance is practiced in the European Alps and in mountainous parts of Asia and Scandinavia.

Transit. Passage of a small object across the face of a larger object, such as a MOON passing across a PLANET.

Transpiration. Loss of moisture to the ATMOSPHERE through the leaves of plants. When considered together with EVAPORATION, the term EVAPOTRANSPIRATION is used.

Transportation. Movement of goods or people from one PLACE to another. In earlier times, and in some poor countries today, animals provide the means of transportation. Most countries now have mechanical transportation, such as trains, buses, automobiles, airplanes, and steamships. Improvements in transport and COMMUNICATIONS, especially in the nineteenth and twentieth centuries, led to what geographers call TIME-SPACE CONVERGENCE. Places were connected more quickly, easily, and cheaply, which contributed to globalization. Transportation based on the burning of FOSSIL FUELS is a major cause of GREENHOUSE GASES.

Transverse bar. Flat-topped body of SAND or gravel oriented transverse to the RIVER flow.

Transverse dunes. Asymmetrical SAND DUNES running at right angles to the prevailing WIND direction. They form where there is an abundant supply of sand and only moderate winds.

Transverse valley. River-cut VALLEY or GORGE that runs perpendicular to the main STRIKE direction of a mountain CHAIN.

Transverse waves. See S WAVES.

Travertine. LIMESTONE formations such as STALACTITES and STALAGMITES that form in limestone CAVES and around calcareous SPRINGS. Also known as TUFA.

Travertine formation, the Liberty Cap, at Mammoth Hot Springs in Yellowstone National Park. (U.S. Geological Survey)

Treeline. See TIMBERLINE.

Trench. Long, deep shape in the OCEAN floor, close to a CONTINENT or an ISLAND ARC. Trenches are formed as part of SUBDUCTION, when OCEANIC CRUST is forced down beneath an adjacent TECTONIC PLATE. Adjacent to a trench is a zone of active VOLCANOES, formed by the heat, pressure, and melting of the descending material. The lowest PLACE on Earth is at the bottom of the Mariana Trench, more than 36,000 feet (11,000 meters) below MEAN SEA LEVEL.

Triassic. PERIOD of time about 225 to 195 million years ago at the beginning of the MESOZOIC ERA when dinosaurs lived.

Tributary. STREAM that joins its water with a larger stream. The smallest tributaries are tiny streams, numbered as first-order streams in a network (see STREAM ORDER). Some tributaries are themselves major RIVERS, such as the Missouri, a tributary of the Mississippi.

Trophic level. Different types of food relations that are found within an ECOSYSTEM. Organisms that derive food and ENERGY through PHOTO-SYNTHESIS are called autotrophs (self-feeders) or producers. Organisms that rely on producers as their source of energy are called heterotrophs (feeders on others) or consumers. A third trophic level is represented by the organisms known as decomposers, which recycle organic waste.

Tropical cyclone. STORM that forms over tropical OCEANS and is characterized by extreme amounts of rain, a central area of calm AIR, and spinning WINDS that attain speeds of up to 180 miles (300 km.) per hour.

Tropical depression. STORM with WIND speeds up to 38 miles (64 km.) per hour.

Tropical rain forest. See RAIN FOREST, TROPICAL.

Tropical storm. STORM with WINDS of 38-70 miles (64-118 km.) per hour.

Tropics. The REGION of the earth lying between the tropic of CAPRICORN, 23.5 DEGREES south, and the tropic of CANCER, 23.5 degrees north. More than one-third of the earth's land lies in the Tropics, with CLIMATES ranging from the hot humid tropical RAIN FOREST to the hot arid tropical DESERT. TEMPERATURES in the Tropics are high all year, because the SUN is always nearly vertically overhead. The annual RANGE of temperature is 77-82 degrees Fahrenheit (25 to 28 degrees Celsius). SEASONS are not measured by temperature variation or by changes in length of DAY, but by the season of RAINFALL (except in the tropical rain forest, where it rains all year). The most spectacular climate change in the Tropics is the MONSOON; the dramatic onset of the Asian monsoon is both eagerly awaited and dreaded. The ancient Greeks believed that human life was not possible in the Tropics, because of the high temperatures, but today the tropical region of Southeast Asia contains about one-fifth of the world's POPULATION.

Tropopause. BOUNDARY layer between the TROPOSPHERE and the STRATO-SPHERE.

Troposphere. Lowest and densest of Earth's atmospheric layers, marked by considerable TURBULENCE and a decrease in TEMPERATURE with increasing ALTITUDE.

Trough. Long, relatively gentle-sided depression or furrow, sometimes subdivided into many smaller troughs.

True horizon. See HORIZON, TRUE.

Tsunami. SEISMIC SEA WAVE caused by a disturbance of the OCEAN floor, usually an EARTHQUAKE, although undersea LANDSLIDES or volcanic ERUPTIONS can also trigger tsunami. A tsunami travels through the ocean at great speed; it has a small WAVE HEIGHT but long WAVE LENGTH. The Japanese word means "HARBOR wave," because when the tsunami reaches the COAST it grows tall and creates tremendous destruction. Tsunami have caused such destruction and loss of life that there is a warning system in place covering the Pacific Ocean, with stations in Alaska and Hawaii ready to transmit instant warning of impending tsunami. It is incorrect to use the term "tidal wave" for a tsunami, since TIDES have nothing to do with tsunami.

Tsunami warning. Second phase of a TSUNAMI alert; it is issued after the generation of a tsunami has been confirmed.

Tsunami watch. First phase of a TSUNAMI alert; it is issued after a large EARTHQUAKE has occurred at the seafloor.

Tufa. LIMESTONE or calcium carbonate deposit formed by PRECIPITATION from an alkaline LAKE. Mono Lake is famous for the dramatic tufa towers exposed by the lowering of the level of lake water. Also known as TRAVERTINE.

Tufa tower on California's Mono Lake. (Corbis)

Tuff. Compacted deposit that is 50 percent or more VOLCANIC ASH and DUST.

Tuff ring. Larger form of a MAAR.

Tumescence. Local swelling of the ground that commonly occurs when MAGMA rises toward the surface.

Tundra. The treeless far northern lands of Canada and Eurasia, covering about one-tenth of the earth's lands. There, the CLIMATE is so cold that only low plants can grow. The ground is snow-covered for eight or more months a year and PRECIPITATION is low. Plants of the tundra are adapted to a short growing season. The Inuit peoples of North America and the Saami of nothern Europe are native to the ARCTIC tundra. In high MOUNTAINS there is a second type of tundra, known as ALPINE tundra. TEMPERATURES there are too low for trees to grow, snow covers the ground for much of the year, and strong WINDS are frequent. Both types of tundra are fragile environments where development for mining, oil drilling, or even recreation threatens the ecosystems.

Tunnel vent. Central tube in a volcanic structure through which material from the earth's interior travels.

Turbulence. Rapid flow of water in rivers, in estuaries, and near ocean surfaces, and the movement of AIR in STORMS. High-speed WINDS and large-scale atmospheric phenomena usually create large differences in fluid velocity over relatively small distances. These highly sheared flows tend to be generically unstable when their otherwise smooth "laminar" motion is subjected to naturally occurring disturbances. The resulting oscillations in air or water velocity tend to grow rapidly in amplitude and can produce a chaotic, highly fluctuating state of fluid motion known as hydrodynamic turbulence. The turbulent motion of fluids is a universal phenomenon, occurring in a wide variety of environmental fluid flows, in the flow of air about aircraft and water about ship and submarine hulls, in the interior motions of stars, and in galactic jets and CLOUDS.

Typhoon. Name used for a HURRICANE or TROPICAL CYCLONE occurring in East Asia, in the East China Sea, and as far north as southern Japan.

U-shaped valley. Steep-sided VALLEY carved out by a GLACIER. Also called a glacial TROUGH.

U-shaped valley in England's Lake District. Erosion is particularly rapid at the heads of glaciers, which press rock fragments against the sides of the valleys they move through, widening and deepening the valleys by abrasion. (Ray Sumner)

Ubac slope. Shady side of a MOUNTAIN, where local or microclimatic conditions permit lower TIMBERLINES and lower SNOW LINES than occur on a sunny side.

Ultimate base level. Level to which a STREAM can erode its bed. For most RIVERS, this is SEA LEVEL. For streams that flow into a LAKE, the ultimate base level is the level of the lakebed.

Ultramafic rocks. Dense, dark-colored, iron- and magnesium-rich silicate ROCKS composed primarily of the MINERALS olivine and pyroxene. They are the dominant rocks in the earth's MANTLE but also occur in some areas of the CRUST. Ultramafic rocks are important for what they contribute to the understanding of crust and mantle evolution. They also serve as an important source of economic commodities such as chromium, platinum, nickel, and diamonds, as well as talc and various decorative building stones.

Ultraviolet radiation. Electromagnetic radiation extending from just above the sensitivity of the human eye. Form of ENERGY that can cause chemical reactions; it has more energy than visible light and contributes to the breakdown of OZONE in Earth's ATMOSPHERE.

Unconfined aquifer. AQUIFER whose upper BOUNDARY is the WATER TABLE; it is also called a water table aquifer.

Unconformity. Interruption or break in the depositional sequence of SEDIMENTARY ROCKS, representing a long PERIOD of geologic time. This might result from a fall in SEA LEVEL or a tectonic event. Generally, an unconformity represents an erosional surface. Other SEDIMENTS later were deposited on top of this erosional surface. Near the bottom of the Grand Canyon walls lies the Great Unconformity.

This small mesa near San Lorenzo Arroyo in Arizona shows an unconformity overlying upthrust rock. (U.S. Geological Survey)

Underclass. Group of people who experience a form of poverty that keeps them isolated from the mainstream POPULATION and from the formal labor market. In the United States, persons with limited English-speaking skills, single parents, and the long-term unemployed are commonly members of the underclass. The underclass is also subject to increased levels of violence and higher-than-average levels of drug use, illness, and crime.

Underemployment. Phenomenon that occurs when people work less than they wish or less than full time. This is often a way to prevent some workers being fired or laid off.

Underfit stream. STREAM that appears to be too small to have eroded the VALLEY in which it flows. A RIVER flowing in a glaciated valley is a good example of underfit.

Uniformitarianism. Theory introduced in the early nineteenth century to explain geologic processes. It used to be believed that the earth was only a few thousand years old, so the creation of LANDFORMS would have been rapid, even catastrophic. This theory, called CATASTROPHISM, explained most landforms as the result of the Great Flood of the Bible, when Noah, his family, and animals survived the deluge. Uniformitarianism, in contrast, stated that the processes in operation today are slow, so the earth must be immensely older than a mere few thousand years.

Universal time (UT). See GREENWICH MEAN TIME.

Universal Transverse Mercator. Projection in which the earth is divided into sixty zones, each six DEGREES of LONGITUDE wide. In a traditional Mercator projection, the earth is seen as a sphere with a cylinder wrapped around the EQUATOR. UTM can be visualized as a series of six-degree side strips running transverse, or north-south.

Universalizing religion. Proselytic RELIGION; one that actively seeks to convert others to its belief system. Followers of a universalizing religion believe that their religion is appropriate for everyone. See also ETHNIC RELIGION.

Unstable air. Condition that occurs when the AIR above rising air is unusually cool so that the rising air is warmer and accelerates upward.

Upland. Land that is higher than nearby RIVER VALLEYS. A PLATEAU is one kind of upland.

Uplift. Rising of the earth's surface or the increase in distance between the earth's surface and its center.

Upper mantle. Comparatively rigid part of the earth's interior below the CRUST of the earth down to about 700 kilometers, composed of magnesium and iron-rich ROCK.

Upthrown block. When EARTHQUAKE motion produces a FAULT, the block of land on one side is displaced vertically relative to the other. The higher is the upthrown block; the lower is the downthrown block.

Upwelling. OCEAN phenomenon in which warm SURFACE WATERS are pushed away from the COAST and are replaced by cold waters that carry more nutrients up from depth.

Urban area. In many PLACES, a SETTLEMENT with two thousand or more residents is considered urban; smaller settlements are RURAL. Generally the POPULATION of an urban area is engaged in secondary or TERTIARY economic activity. A large urban settlement is a CITY or MEGALOPOLIS.

Urban heat island. Cities experience a different MICROCLIMATE from surrounding REGIONS. The CITY TEMPERATURE is typically higher by a few DEGREES, both DAY and night, because of factors such as surfaces with higher heat absorption, decreased WIND strength, human heat-producing activities such as power generation, and the layer of AIR POLLUTION (DUST DOME).

Urbanization. Increase in the proportion of a POPULATION living in URBAN AREAS.

UTC. See COORDINATED UNIVERSAL TIME.

UTM. See UNIVERSAL TRANSVERSE MERCATOR.

Uvala. Slavic term for an enlarged SINKHOLE in LIMESTONE, or a KARST VALLEY.

Vadose zone. The part of the SOIL also known as the zone of AERATION, located above the WATER TABLE, where space between particles contains AIR.

Valley. Natural LANDFORM in which a long low shape is surrounded by higher valley sides reaching up to a valley crest. Valleys are eroded over

Valleys are natural landforms in which long, low shapes are surrounded by higher valley sides reaching up to crests. Valleys are eroded over time by streams, or sometimes by glaciers. (PhotoDisc)

time by STREAMS, or sometimes by GLACIERS. RIFT VALLEYS are created by tectonic movement. RIVER VALLEYS have been the hearths of many great CULTURES of the past, such as the Egyptians on the Nile River.

Valley glacier. See ALPINE GLACIER.

Valley train. Fan-shaped deposit of glacial MORAINE that has been moved down-valley and redeposited by MELTWATER from the GLACIER.

Van Allen radiation belts. Bands of highly energetic, charged particles trapped in Earth's MAGNETIC FIELD. The particles that make up the inner BELT are energetic protons, while the outer belt consists mainly of electrons and is subject to DAY-night variations.

Varnish, desert. Shiny black coating often found over the surface of ROCKS in arid REGIONS. This is a form of OXIDATION or CHEMICAL WEATHERING, in which a coating of manganese oxides has formed over the exposed surface of the rock.

Varve. Pair of contrasting layers of SEDIMENT deposited over one year's time; the summer layer is light, and the winter layer is dark.

Vegetation. The plant life of an area. The four broad types are FOREST, GRASSLANDS, TUNDRA, and DESERT (or XEROPHYTIC). Forests are found in the TROPICS and the midlatitudes, wherever there is sufficient RAINFALL. Grassland dominates in REGIONS of lower rainfall. Tundra vegetation is small in size because of high TEMPERATURES and a permanently frozen ground surface known as PERMAFROST. Desert vegetation is adapted to low PRECIPITATION, less than one foot (30 centimeters) per year. Deserts are found in the Tropics and in midlatitudes.

Vein. MINERAL deposit that fills a crack; veins form by PRECIPITATION of minerals from fluids.

Veld. South African term for GRASSLANDS found when the early Dutch settlers, or Boers, trekked inland onto the PLATEAU. Afrikaans word for "field." Southern Africa's plateau REGION is known as the high veld.

Ventifacts. PEBBLES on which one or more sides have been smoothed and faceted by ABRASION as the WIND has blown SAND particles.

Vernacular region. CULTURE REGION that is identified by both the majority of people living within the region and by people living outside the region. People living in a vernacular region have a sense of regional identity. The vernacular region usually has a name that is widely understood by large numbers of people. Vernacular regions have a regional identity associated with the name. An example of a vernacular region in the United States is Appalachia.

Vernal equinox. See EQUINOX.

Village. Small SETTLEMENT, usually in a RURAL area. URBAN geographers regard a village as being larger than a HAMLET but smaller than a TOWN. Generally, this means that a village has fewer than twenty-five hundred residents, most of whom are engaged in agricultural activity.

Volcanic ash. Also known as volcanic DUST, the fine pyroclastic material thrown into the AIR in explosive volcanic ERUPTIONS. These particles are small enough to be held in SUSPENSION in the ATMOSPHERE, and can be spread around the whole earth by upper-level WINDS. The eruption of the Philippines' Mount Pinatubo in June, 1991, sent about twenty million tons of volcanic ash, dust, and gases into the atmosphere. The resulting cloud was spread by global winds, producing a band around the entire earth, extending from twenty DEGREES north LATITUDE to thirty degrees north latitude. Because of the increased reflection of SOLAR RADIATION, TEMPERATURES were slightly lower than usual for two years following the eruption.

Volcanic earthquakes. Small-magnitude EARTHQUAKES that occur at relatively shallow depths beneath active or potentially active VOLCANOES.

Volcanic island arc. Curving or linear group of volcanic ISLANDS associated with a SUBDUCTION ZONE. See also OCEANIC ISLANDS.

Volcanic neck. The throat of a VOLCANO, or the pipelike opening in which the LAVA rises up before an explosion. Sometimes the lava solidifies inside the opening, then the surrounding cone is eroded away, leaving the neck exposed as a tall, steep-sided LANDFORM. Ship Rock, New Mexico, and Devil's Tower, Wyoming, are well-known volcanic necks.

Volcanic plumes. Material thrown up from the surface by ERUPTIONS; they indicate high volcanic activity.

Volcanic rock. Type of IGNEOUS ROCK that is erupted at the surface of the earth; volcanic rocks are usually composed of larger crystals inside a fine-grained matrix of very small crystals and glass.

Volcanic tremor. Continuous vibration of long duration, detected only at active VOLCANOES.

Volcano. Geologic phenomenon produced by the ERUPTION of MAGMA from beneath a PLANET's surface; it creates MOUNTAINS that often display a cone shape. Volcanic activity usually occurs in SUBDUCTION ZONES, in SEAFLOOR SPREADING zones, and at HOT SPOTS. When one TECTONIC PLATE is pushed or dragged beneath another plate, the pro-

Oregon's Mount Hood, part of the Pacific Coast's Cascade Range, is a beautiful example of a cone-shaped volcano. As is the case with Japan's Mount Fuji, Mount Hood's appearance is enhanced by the absence other nearby mountains. (PhotoDisc)

cess is called SUBDUCTION. Friction melts the descending ROCK, which is also full of steam, and the hot magma rises until it bursts through the overlying plate in a volcano. When plates diverge at a seafloor spreading zone, magma spreads out from the RIFT, forming long MID-OCEAN RIDGES of volcanoes. Hot spots, or MANTLE PLUMES, are small REGIONS of the earth's CRUST where magma rises in a thin stream to the surface, as in Hawaii, or close to the surface, as in Yellowstone National Park. Volcanic cones are classified by their composition into CINDER CONES, SHIELD VOLCANOES, and COMPOSITE CONES or STRATOVOLCANOES.

Volcanology. Scientific study of VOLCANOES.

Voluntary migration. Movement of people who decide freely to move their place of permanent residence. It results from PULL FACTORS at the chosen destination, together with PUSH FACTORS in the home situation.

Wadi. Arabic word for a WASH, or dry STREAMBED.

Warm cloud. Visible SUSPENSION of tiny water droplets at TEMPERATURES above freezing.

Warm front. See FRONT.

Warm temperate glacier. GLACIER that is at the melting TEMPERATURE throughout.

Wash. Dry STREAMBED, filled with ALLUVIUM, in an arid area. ARROYO is the Spanish word for this feature.

Water cycle. Continuous movement of the water of the earth's HYDRO-SPHERE. EVAPORATION from the OCEAN is followed by CONDENSATION into CLOUDS, then PRECIPITATION. RUNOFF returns water to the ocean, where it is again evaporated. Also called the HYDROLOGIC CYCLE.

Water gap. Low point in a RIDGE through which a STREAM flows. Generally, a water gap indicates UPLIFT of the REGION while the stream has continued to erode its bed.

Water power. Generally means the generation of electricity using the EN-ERGY of falling water. Usually a DAM is constructed on a RIVER to provide the necessary height difference. The potential energy of the falling water is converted by a water turbine into mechanical energy. This is used to power a generator, which produces electricity. Also called HYDROELECTRIC POWER. Another form of water power is tidal power, which uses the force of the incoming and outgoing TIDE as its source of energy.

Water resources. All the SURFACE WATER and GROUNDWATER that can be effectively harvested by humans for domestic, industrial, or agricultural uses.

Water table. The depth below the surface where the zone of AERATION meets the zone of SATURATION. Above the water table, there may be some SOIL MOISTURE, but most of the pore space is filled with air. Below the water table, pore space of the ROCKS is occupied by water that has percolated down through the overlying earth material. This water is called GROUNDWATER. In practice, the water table is rarely as flat as a table, but curved, being far below the surface in some PLACES and even intersecting the surface in others. When GROUNDWATER emerges at the surface, because it intersects the water table, this is called a SPRING. The depth of the water table varies from SEASON to season, and with pumping of water from an AQUIFER.

Waterfall. Part of a STREAM where there is a steep, nearly perpendicular, fall in the STREAMBED. Waterfalls often form where there is resistant ROCK in one part of the streambed and softer rock in the next section. The softer rock is eroded more rapidly, leaving a rock ledge over which the water then falls. Other waterfalls are caused by EARTHQUAKE faulting or by stream REJUVENATION due to UPLIFT. Over time, the edge of the waterfall recedes because of EROSION of the lip. Another term for a waterfall is a KNICKPOINT.

Oregon's Multnomah Falls. (PhotoDisc)

Watershed. The whole surface area of land from which RAINFALL flows downslope into a STREAM. The watershed comprises the STREAMBED or CHANNEL, together with the VALLEY sides, extending up to the crest or INTERFLUVE, which separates that watershed from its neighbor. Each watershed is separated from the next by the drainage DIVIDE. Also called a DRAINAGE BASIN.

Waterspout. TORNADO that forms over water, or a tornado formed over land which then moves over water. The typical FUNNEL CLOUD, which reaches down from a CUMULONIMBUS CLOUD, is a narrow rotating STORM, with WIND speeds reaching hundreds of miles per hour.

Wave. Moving SWELL on the surface of a body of water. In the deep OCEAN are waves of oscillation (waves of transition), where the wave ENERGY, but not the water, is moving forward. The friction of WINDS blowing over the surface of the ocean creates waves. As ocean waves approach the SHORE and water becomes shallower, they change to waves of translation, in which the water and the energy both move toward the shore. See also BREAKER.

Wave crest. Top of a WAVE.

Wave-cut platform. As SEA CLIFFS are eroded and worn back by WAVE attack, a wave-cut platform is created at the base of the cliffs. ABRASION by ROCK debris from the cliffs scours the platform further, as waves wash to and fro and TIDES ebb and flow. The upper part of the wave-cut platform is exposed at high tide. These areas contain rockpools, which are rich in interesting MARINE life-forms. Offshore beyond the platform, a wave-built TERRACE is formed by DEPOSITION.

Wave height. Vertical distance between one WAVE CREST and the adjacent WAVE TROUGH.

Wave length. Distance between two successive WAVE CRESTS or two successive WAVE TROUGHS.

Wave trough. The low part of a WAVE, between two WAVE CRESTS.

Weather. DAY-to-day variations in atmospheric conditions, including TEMPERATURE, precipitation, humidity, cloud cover, winds or STORMS, and ATMOSPHERIC PRESSURE conditions. Weather is constantly changing, and scientists study it so as to make predictions or forecasts. CLIMATE is the long-term average of recorded weather data.

Weather analogue. Approach to WEATHER FORECASTING that uses the WEATHER behavior of the past to predict what a current weather pattern will do in the future.

Weather forecasting. Attempt to predict WEATHER patterns by analysis of current and past data.

Weathering. The change or breaking down of ROCK when it is exposed at the earth's surface. PHYSICAL WEATHERING is the breaking down into smaller pieces, or disintegration; CHEMICAL WEATHERING is the process of decomposition through chemical change. Weathering is a prelude to EROSION.

Well. Artificial entry into the WATER TABLE. Both farmers and cities sink wells to tap GROUNDWATER.

Western Hemisphere. The half of the earth containing North and South America; generally understood to fall between LONGITUDES 160 DE-GREES east and 20 degrees west.

Wetlands. PLACES where the ground is saturated with water, Specialized VEGETATION, called hygrophytic plants, grows there. Wetlands are a transition between aquatic ECOSYSTEMS and terrestrial ecosystems. COASTLINES where DEPOSITION is occurring commonly have wetlands in estuaries and infilled LAGOONS. These wetlands are classed as SALT MARSH. Tropical COASTAL WETLANDS have MANGROVE SWAMPS. There can be wetlands with FRESH WATER, as in PEAT BOGS in northern LATI-TUDES or backswamps on FLOODPLAINS. Wetlands are rich biological reservoirs but are greatly endangered by DEVELOPMENT.

Wetlands are places in which the ground is permanently saturated with water. Wetlands form transitions between aquatic and terrestrial ecosystems. (Photo-Disc)

White water. Turbulent and frothy portions of RAPIDS; so called because of the tendency of the water to form white foam.

Wilderness. Originally, a PLACE where no humans lived, generally be-cause of harsh conditions. In the second half of the twentieth century, as roads were constructed into MOUNTAINS and FORESTS and off-road vehicles became widely used for recreation, people came to see a need for preservation of parts of the COUNTRY that would only be accessible on foot—wilderness areas. The Wilderness Act of 1964 defined "wil-derness" as "a place that is not controlled by humans, where natural ecological processes operate freely and where its primeval character and influences are retained; a place that is not occupied or modified by mankind, where humans are visitors, and the imprint of their activity is largely unnoticeable; a place with outstanding opportunities for the solitude necessary for a primitive and unconfined recreation experience." Numerous areas have been set aside as wilderness areas in NATIONAL PARKS, in national forests, and on land controlled by the Bureau of Land Management. In the year 2000, there were forty-nine national park wilderness areas in the United States.

Willy willy. Australian term for a DUST DEVIL. (Americans sometimes mistakenly believe it is an Australian term for HURRICANE.)

Wilson cycle. Creation and destruction of an OCEAN BASIN through the process of SEAFLOOR SPREADING and SUBDUCTION of existing ocean basins.

Wind. Horizontal movement of AIR relative to the earth's surface, caused by differences in ATMOSPHERIC PRESSURE. These pressure differences arise largely because of unequal heating of the earth's surface by the SUN's rays. Winds play an important role in WEATHER and CLIMATE. Winds that blow predominantly from one direction are called prevailing winds. Before the invention of steamships, sailing ships relied on prevailing winds to cross OCEANS.

Wind energy. Power generated using the force of the WIND. Windmills have provided wind energy for centuries, generally to pump water or grind grains. In the late twentieth century, concern about the use of FOSSIL FUELS led to research into generating power through wind energy, which is a RENEWABLE and nonpolluting source of energy. California has been a world leader in modern wind-generation TECHNOL-

This array of wind turbines at Altamont in Northern California is part of one of the largest arrays of wind turbines in the United States. (PhotoDisc)

ogy, with major wind farms located near Palm Springs. Although this is a sustainable energy form, some people are opposed to the appearance of fields of wind generators.

Wind gap. Abandoned WATER GAP. The Appalachian Mountains contain both wind gaps and water gaps.

Windbreak. Barrier constructed at right angles to the prevailing WIND direction to prevent damage to crops or to shelter buildings. Generally, a row of trees or shrubs is planted to form a windbreak. The feature is also called a shelter belt.

Windchill. MEASUREMENT of apparent TEMPERATURE that quantifies the effects of ambient WIND and temperature on the rate of cooling of the human body.

Windward. Front or exposed side of a MOUNTAIN or RANGE is the windward side. A RAIN SHADOW is usually located on the windward side of a mountain range. Compare to LEEWARD.

Winter solstice. DAY on which winter begins; in the NORTHERN HEMISPHERE, about December 21, and in the SOUTHERN HEMISPHERE, about June 21.

Woodlands. VEGETATION communities in which the upper canopy is not completely closed because the trees are more widely spaced than in a FOREST. In the intervening spaces, shrubs and other groundcover grow. Woodland is a response to drier conditions, where RAINFALL is not sufficient for true forest to grow. SAVANNA woodland grades into savanna GRASSLAND with increasing ARIDITY.

Woodlands differ from forests in having upper canopies that are not completely closed to sunlight because their trees are more widely spaced than those in forests. (PhotoDisc)

World Aeronautical Chart. International project undertaken to map the entire world, begun during World War II.

World Bank. Bank providing developmental assistance in the form of financial loans to client countries. The World Bank and the International Monetary Fund (IMF) were created after World War II, following the Bretton Woods Conference of 1944. There were 160 members of the World Bank in 2000.

World city. CITY in which an extremely large part of the world's economic, political, and cultural activity occurs. In the year 2000, the three world cities were London, New York, and Tokyo.

Xenolith. Smaller piece of ROCK that has become embedded in an IGNEOUS ROCK during its formation. It is a piece of older rock that was incorporated into the fluid MAGMA.

Xeric. Description of SOILS in REGIONS with a MEDITERRANEAN CLIMATE, with moist cool winters and long, warm, dry summers. Since summer is the time when most plants grow, the lack of SOIL MOISTURE is a limiting factor on plant growth in a xeric ENVIRONMENT.

Xerophytic plants. Plants adapted to arid conditions with low PRECIPITATION. Adaptations include storage of moisture in tissue, as with cactus plants; long taproots reaching down to the WATER TABLE, as with DESERT shrubs; or tiny leaves that restrict TRANSPIRATION.

Yardangs. Small LANDFORMS produced by WIND EROSION. They are a series of sharp RIDGES, aligned in the direction of the wind.

Yazoo stream. TRIBUTARY that flows parallel to the main STREAM across the FLOODPLAIN for a considerable distance before joining that stream. This occurs because the main stream has built up NATURAL LEVEES through flooding, and because RELIEF is low on the floodplain. The yazoo stream flows in a low-lying wet area called backswamps. Named after the Yazoo River, a tributary of the Mississippi.

Zero population growth. Phenomenon that occurs when the number of deaths plus EMIGRATION is matched by the number of births plus IMMIGRATION. Some European countries have reached zero population growth.

Zone of ablation. See ABLATION, ZONE OF.

Zone of accumulation. See ACCUMULATION, ZONE OF.

Zone of aeration. See AERATION, ZONE OF.

Zone of saturation. See SATURATION, ZONE OF.

Zoning. Land-management tool used to limit uses and define conditions and extent of use.

BIBLIOGRAPHY

THE NATURE OF GEOGRAPHY

Harley, J. B., and David Woodward, eds. *The History of Cartography: Cartography in the Traditional Islamic and South Asian Societies.* Vol. 2, book 1. Chicago: University of Chicago Press, 1992. Offers a critical look at maps, mapping, and mapmakers in the Islamic world and South Asia.
_____, eds. *The History of Cartography: Cartography in the Traditional East and Southeast Asian Societies.* Vol. 2, book 2. Chicago: University of Chicago Press, 1994. Similar in thrust and breadth to volume 2, book 1.
Woodward, David, et al., eds. *The History of Cartography: Cartography in the Traditional African, American, Arctic, Australian, and Pacific Societies.* Vol. 2, book 3. Chicago: University of Chicago Press, 1998. Investigates the roles that maps have played in the wayfinding, politics, and religions of diverse societies such as those in the Andes, the Trobriand Islanders of Papua-New Guinea, the Luba of central Africa, and the Mixtecs of Central America.
Woodward, David, and J. B. Harley, eds. *The History of Cartography: Cartography in Prehistoric, Ancient and Medieval Europe and the Mediterranean.* Vol. 1. Chicago: University of Chicago Press, 1987. Critical look at early European and Mediterranean mapmaking.

PHYSICAL GEOGRAPHY

Lutgens, Frederick K., and Edward J. Tarbuck. *Foundations of Earth Science.* Upper Saddle River, N.J.: Prentice-Hall, 1998. Undergraduate text for an introductory course in earth science, consisting of seven units covering basic principles in geology, oceanography, meteorology, and astronomy, for those with little background in science.
McKnight, Tom. *Physical Geography: A Landscape Appreciation.* 6th ed. New York: Prentice Hall, 2000. Now classic college textbook that has become popular because of its illustrations, clarity, and wit. Comes with a CD-ROM that takes readers on virtual-reality field trips.
Robinson, Andrew. *Earth Shock: Climate Complexity and the Force of Nature.* New York: W. W. Norton, 1993. Describes, illustrates, and analyzes the forces of nature responsible for earthquakes, volcanoes, hurricanes, floods, glaciers, deserts, and drought. Also recounts how humans have perceived their relationship with these phenomena throughout history.
Strahler, Alan, and Arthur Strahler. *Physical Geography.* 2d ed. New York: Wiley, 2001. A popular introductory physical geography textbook containing a readable account of the world's climates.
Weigel, Marlene. *UxL Encyclopedia of Biomes.* Farmington Hills, Mich.: Gale Group, 1999. This three-volume set should meet the needs of seventh grade classes for research. Covers all biomes such as the forest, grasslands, and desert. Each biome includes sections on development of that particular biome, type, and climate, geography, and plant and animal life.

HUMAN GEOGRAPHY

Erickson, Jon. *The Human Volcano: Population Growth as Geologic Force.* New York: Facts on File, 1995. Reveals the geographic effects of overpopulation on planetary resources such as wildlife habitats, food availability, climatic conditions, and agriculture. It discusses the human impact on the Earth's natural cycles and introduces the concept of carrying capacity.

Glantz, Michael H. *Currents of Change: El Niño's Impact on Climate and Society.* New York: Cambridge University Press, 1996. Aids readers in understanding the complexities of the earth's weather pattern, how it relates to El Niño, and the impact upon people around the globe.

Hunter, Malcolm L., Jr. *Fundamentals of Conservation Biology.* 2d ed. Malden, Mass.: Blackwell Science, 2002. Introduces and explains the concept of conservation biology and the applied science of maintaining the earth's biological diversity. Addresses social, political, and economic issues in a manner that can be readily understood by people outside of the field who are concerned about the future of Earth and its inhabitants.

Novaresio, Paolo. *The Explorers: From the Ancient World to the Present.* New York: Stewart, Tabori and Chang, 1996. Describes amazing journeys and exhilarating discoveries from the earliest days of seafaring to the first landing on the moon and beyond.

Reid, T. R. "Feeding the Planet." *National Geographic* 194, no. 4 (October, 1998): 56-75. So far, global food production has kept pace with a burgeoning population. Maintaining that balance and finding ways to share Earth's bounty are critical challenges.

ECONOMIC GEOGRAPHY

Chandler, Gary, and Kevin Graham. *Alternative Energy Sources.* Breckinridge, Colo.: Twenty First Century Books, 1996. Geared for young adults, this volume focuses on the types of resources that will not damage the environment.

Esping-Andersen, Gosta. *Social Foundations of Postindustrial Economies.* New York: Cambridge University Press, 1999. Examines such topics as social risks and welfare states, the structural bases of postindustrial employment, and recasting welfare regimes for a postindustrial era.

Prevost, P., and P. Le Gloru. *Fundamentals of Modern Agriculture.* Enfield, N.H.: Science Publishers, 1997. Includes chapters entitled "Present-day Agriculture," "The Agricultural Farm: A Global Approach," "A Cultivated Plant," and "Post-harvest Technology."

Robertson, Noel, and Kenneth Blaxter. *From Dearth to Plenty: The Modern Revolution in Food Production.* New York: Cambridge University Press, 1995. Tells a story of scientific discovery and its exploitation for technological advance in agriculture. It encapsulates the history of an important period, 1936-86, when government policy sought to aid the competitiveness of the agricultural industry through fiscal measures and by encouraging scientific and technical innovation.

REGIONAL GEOGRAPHY

Biger, Gideon, ed. *The Encyclopedia of International Boundaries.* New York: Facts on File, 1995. Entries for approximately two hundred countries are arranged alphabetically, each beginning with introductory information describing demographics, political structure, and political and cultural history. The boundaries of each state are then described with details of the geographical setting, historical background, and present political situation, including unresolved claims and disputes.

NORTH AMERICA AND CARIBBEAN REGION

PHYSICAL GEOGRAPHY

Jones, David. *North American Wildlife.* Portland, Ore.: Graphic Arts Center, 1999. Stunning look at the continent's most amazing creatures, from the vast herds of caribou that roam the Arctic tundra to the reptiles that inhabit Florida's wetlands.

Kricher, John G. *Forests.* Boston: Houghton Mifflin, 1996. Describes and illustrates the fifty different kinds of forest and related habitats found throughout the United States and Canada, from the boreal forest and tundra of the north to the mangrove swamps, desert scrub, and giant saguaro forests of the southwest.

Maingot, Anthony P. *The United States and the Caribbean: Challenges of an Asymmetrical Relationship.* Boulder, Colo.: Westview Press, 1994. Explores the complex interdependence between the small Caribbean states and the United States and looks at their changing relationships throughout history.

Maul, George A., ed. *Climatic Change in the Intra-Americas Sea: Implications of Future Climate on the Structure.* New York: John Wiley and Sons, 1995. Expert and comprehensive account of the implications that global warming and sea level rise will have on the ecosystems and socioeconomic structure in the marine and coastal regions of the Caribbean, Gulf of Mexico, Bahamas, Bermuda, and the northeast coast of South America.

Miller, Ralph L. et al., eds. *Energy and Mineral Potential of the Central American-Caribbean Region.* New York: Springer-Verlag, 1995. Presents studies that examine in detail the energy and mineral resources of Central America and the Caribbean.

Sealey, Neil E. *Caribbean World: A Complete Geography.* London: Macmillan Caribbean, 1992. Provides comprehensive explanations of the natural and human factors that affect the region, from geological and climatological phenomena to population growth.

Wemert, Susan J. *North American Wildlife: An Illustrated Guide to 2,000 Plants and Animals.* Pleasantville, N.Y.: Reader's Digest Association, 1998. Spans the land from Florida to the Northwest Territories and embraces fields, forests, ponds, and prairies and includes more than two thousand plants and animals of all types.

HUMAN GEOGRAPHY

Bean, Frank D., et al., eds. *At the Crossroads: Mexican Migration and U.S. Policy.* Lanham, Md.: Rowman and Littlefield, 1997. Comprehensive collection of chapters such as "Mexico and U.S. Worldwide Immigration Policy," "Mexican Immigration and the U.S. Population," and "Fiscal Impacts of Mexican Migration to the United States."

Castaneda, Jorge G. *The Mexican Shock: Its Meaning for the United States.* New York: New Press, 1995. Provides a vision of the meaning of developments for the future such as NAFTA and California's "Proposition 187" on immigration, the collapse of the peso and the subsequent U.S. bailout, the uprising in Chiapas, and the unresolved assassination of Mexican presidential candidate Luis Donaldo Colosio.

Collier, Christopher, and James Lincoln Collier. *Hispanic America, Texas, and the Mexican War 1835-1850.* Tarrytown, N.Y.: Marshall Cavendish, 1998. Examines the settlement of the area that became the southwestern portion of the United States, detailing how it evolved from land settled by Native Americans, to Spanish territory, to states that were pawns between the North and the South prior to the Civil War.

Monge, Jose Trias. *Puerto Rico: The Trials of the Oldest Colony in the World.* New Haven, Conn.: Yale University Press, 1999. The island of Puerto Rico has a severely distressed economy, is one of the most densely populated places on earth, and enjoys only limited political freedom. In this book a distinguished Puerto Rican legal scholar and former government official discusses the island's century-old relationship with the United States and argues that the process of decolonization should begin immediately.

Needler, Martin C. *Mexican Politics.* Westport, Conn.: Praeger, 1995. Examines the postrevolutionary Mexican political system and the political and economic influences that are transforming it.

Patterson, Thomas G. *Contesting Castro: The United States and the Triumph of the Cuban Revolution.* New York: Oxford University Press, 1995. Story of Cuban dictator Fulgencio Batista's fall, Castro's triumph, and the roots of Cuban-American enmity lays bare the failures of U.S. policy.

Salutin, Rick. *1837: William Lyon MacKenzie and the Canadian Revolution.* Toronto: Theatre Communications Group (Playwrights Canada Press), 1998. Lively, humorous, and ultimately tragic look at Canada's ill-starred revolution for national independence.

Smith, David E. *The Invisible Crown: The First Principle of Canadian Government.* Toronto: University of Toronto Press, 1995. Presents a perspective on the Crown in Canadian politics as a structuring principle of government. He traces Canada's distinctive form of federalism—with highly autonomous provinces—to the influence of the Crown, going so far as to characterize Canadian government as a system of compound monarchies.

Stephens, Sonya, and Richard Steckel. *A Population History of North America.* New York: Cambridge University Press, 2000. Describes the peopling of North America by the various immigrant groups, the demographics, each group's percentage of the population, and their subsequent role in society.

ECONOMIC GEOGRAPHY

Ayala, Cesar J. *American Sugar Kingdom: The Plantation Economy of the Spanish Caribbean, 1898û1934.* Chapel Hill: University of North Carolina Press, 1999. Focuses on the development of plantation economies in Cuba, Puerto Rico, and the Dominican Republic in the early twentieth century. It focuses on how closely the development of the Spanish Caribbean's modern economic and social class systems is linked to the history of the U.S. sugar industry during its greatest period of expansion and consolidation.

Cremeans, Jack E., ed. *Handbook of North American Industry.* 2d ed. Lanham, Md.: Bernan Associates, 1999. Presents narrative articles and comparative statistical data on the economies of the three member states of NAFTA: the U.S., Canada, and Mexico.

Pillsbury, Richard, and John Florin. *Atlas of American Agriculture: The American Cornucopia.* New York: Macmillan Library Reference, 1996. The first half discusses twelve agricultural regions (including brief sections on Alaska and Hawaii). The second covers twenty-four specific crops, from aquaculture to wheat.

REGIONAL GEOGRAPHY

Fisher, Ron M., and William R. Gray, eds. *Heartland of a Continent: America's Plains and Prairies.* Washington, D.C.: National Geographic Society, 1994. More than a hundred photographs capture the sweep and the space of America's central grasslands in this witty, touching look at the plains, the prairies, and the people of the continent's heartland.

Gore, Rick. "Cascadia." *National Geographic* 193, no. 5 (May, 1998): 6-37. This article looks at the physical forces and geology of the Pacific Northwest, especially the Cascade Mountain country.

Homberger, Eric. *The Penguin Historical Atlas of North America.* New York: Penguin USA, 1995. Examines the history of North America's three principal nations, the U.S., Canada, and Mexico, from their colonial origins to the formation of the North American Free Trade Association. The survey follows the rise of the U.S. to superpower status and assesses the relation of the three nations as a whole to the rest of the world.

Parfait, Michael. "Mexico, A Special Issue." *National Geographic* 190, no. 2 (August, 1996): 2-131. This entire issue is dedicated to Mexico. The titles of the articles are "Emerging Mexico," "Mexico City," "Sierra Madre," "Monterrey," "Veracruz," "Heartland and the Pacific," "Tijuana and the Border," "Yucatán Peninsula," and "Chiapas."

SOUTH AMERICA AND CENTRAL AMERICA

PHYSICAL GEOGRAPHY

Georges, D. V. *South America.* Danbury, Conn.: Children's Press, 1986. Discusses characteristics of various sections of South America such as the Andes, the Amazon rain forest, and the pampas.

Matthews, Down, and Kevin Schaefer. *Beneath the Canopy: Wildlife of the Latin American Rain Forest.* San Francisco: Chronicle Books, 1999. Kevin Schaefer's photographs offer a rare, up-close look at the beautiful and elusive creatures that make their home in this natural paradise—from its leafy shadows to the forest canopy. Captions and text by nature writer Matthews give further insight into the lives of these amazing animals.

HUMAN GEOGRAPHY

Early, Edwin, et al., eds. *The History Atlas of South America.* Foster City, Calif.: IDG Books Worldwide, 1998. Describes South America's history, which is a rich tapestry of complex ancient civilizations, colonial clashes, and modern growth, economic challenges, and cultural vibrancy.

Kelly, Philip. *Checkerboards and Shatterbelts: The Geopolitics of South America.* Austin: University of Texas Press, 1997. Uses the geographical concepts of "checkerboards" and "shatterbelts" to characterize much of South America's geopolitics and to explain why the continent has never been unified or dominated by a single nation.

Levine, Robert M., and John J. Crocitti, eds. *The Brazil Reader: History, Culture, Politics.* Durham, N.C.: Duke University Press, 1999. Selections range from early colonization to the present day, with sections on imperial and republican Brazil, the days of slavery, the Vargas years, and the more recent return to democracy.

Levinson, David, ed. *The Encyclopedia of World Culture, Vol. 7: South America.* Indianapolis, Ind.: Macmillan, 1994. Addresses the diverse cultures of South America south of Panama, with an emphasis on the American Indian cultures, although the African-American culture and the European and Asian immigrant cultures are also covered. Linguistics, historical and cultural relations, economy, kinship, marriage, sociopolitical organizations, and religious beliefs are among the topics discussed for each culture.

Webster, Donovan. "Orinoco River" *National Geographic* 193, no. 4 (April, 1998): 2-31. Examination of the Orinoco River Basin in the Amazonian portion of Venezuela. The article focuses on the fauna, flora, and the Yanomani, Yekwana, and Piaroa tribes and "tropical cowboys."

ECONOMIC GEOGRAPHY

Biondi-Morra, Brizio. *Hungry Dreams: The Failure of Food Policy in Revolutionary Nicaragua.* Ithaca, N.Y.: Cornell University Press, 1993. Examines how food policy was formulated in Nicaragua and the effects on foreign exchange, food prices, and the relationship to wages and credit.

Wilken, Gene C. *Good Farmers: Traditional Agriculture Resource Management in Mexico and Central America.* Berkeley, Calif.: University of California Press, 1987. Focusing on the farming practices of Mexico and Central America, this book examines in detail the effectiveness of sophisticated traditional methods of soil, water, climate, slope, and space management that rely primarily on human and animal power.

REGIONAL GEOGRAPHY

Edwards, Mike. "El Salvador." *National Geographic* 188, no. 3 (September, 1995): 108-131. An overview of present-day El Salvador and how it has recovered from the internal strife of years past.

Egan, E. W. *Argentina in Pictures.* Minneapolis, Minn.: Lerner Publications, 1994. Introduction to the geography, history, government, people, and economy of the second largest South American country.

Gheerbrandt, Alain. *The Amazon: Past, Present and Future.* New York: Harry N. Abrams, 1992. Presents the past, present, and uncertain future of the Amazon rain forest and its inhabitants. It includes spectacular illustrations and a section of historical documents.

Haverstock, Nathan A. *Uruguay in Pictures.* Minneapolis, Minn.: Lerner, 1987. Introduces the land, history, government, people, and economy of a small South American country.

McCarry, John. "Suriname." *National Geographic* 197, no. 6 (June, 2000): 38-55. An overview of Suriname, which is home to a spectrum of native, Asian, African, and European cultures struggling to build a common future.

Putman, John J. "Cuba." *National Geographic* 195, no. 6 (June, 1999): 2-35. Cuba's revolution ages, perhaps mellows, but keeps its grip on this island nation. There is recognition, however, of the growing power of the U.S. dollar in Cuba.

Sumwait, Martha Murray. *Ecuador in Pictures.* Minneapolis, Minn.: Lerner Publications, 1987. Text and photographs introduce the geography, history, economy, culture, and people of the South American country whose name derives from the equator.

AFRICA

PHYSICAL GEOGRAPHY

Chadwick, Douglas H. "Elephants—Out of Time, Out of Space." *National Geographic* 179, no. 5 (May, 1991): 2-49. An extensive article that does a thorough survey of elephants and their problems. It includes statistics on elephant populations in every African and Asian country in 1989.

Conniff, Richard. "Africa's Wild Dogs." *National Geographic* 195, no. 5 (May, 1999): 36-63. These parti-color canines, one of Africa's most endangered species, are not as well known as lions or leopards but demonstrate fascinating social behavior. The narration follows a number of packs of dogs the authors studied for several years in the Okavango Delta region of Botswana.

Disilvestro, Roger L., ed. *The African Elephant: Twilight in Eden.* New York: John Wiley & Sons, 1991. Discusses elephant evolution and biology, modern ivory poaching and trade, recent conservation efforts, and finally the problems that elephants face today.

Estes, Richard Despard. *The Behavior Guide to African Mammals: Including Hoofed Mammals, Carnivores, Primates.* Berkeley: University of California Press, 1991. Describes and explains the behavior of four major groups of mammals.

James, Valentine Udoh. *Africa's Ecology: Sustaining the Biological and Environmental Diversity of a Continent.* Jefferson, N.C.: McFarland, 1993. Case studies and photographs document the effects of a growing demand for consumer goods. They consider parks and reserves, forests, savannas, deserts, and water resources; the impact of tourism, agriculture, and other activities; and the role of women in protecting resources.

Karekezi, Stephen, and Timothy Ranja. *Renewable Energy Technologies in Africa.* New York: St. Martin's Press, 1997. Sums up the whole of Eastern Africa (including the Horn) and Southern Africa (including South Africa itself) with regard to what is known about the innovations and deployment of renewable energy technologies in the regions.

Linden, Eugene. "Bonobos, Chimpanzees With a Difference." *National Geographic* 181, no. 2 (March, 1992): 46-53. A short study of a separate species in the Congo (Zaire) which some say is our closest relative among all of the primates.

Harcourt, Caroline S., and Jeffrey A. Sayer. *The Conservation Atlas of Tropical Forests: Africa.* Upper Saddle River, N.J.: Prentice-Hall, 1992. Part 1 of this volume presents an overview of Africa's tropical forests. Topics discussed include history; biological diversity; effects of population, agriculture, and the timber trade; forest management and conservation; and comments concerning the future. Twenty-two chapters in Part 2 deal with individual African countries or a small group of related countries.

Steentoft, Margaret. *Flowering Plants in West Africa*. New York: Cambridge University Press, 1988. Professor Steentoft provides an account of the flowering plant flora of West Africa south of the Sahara (Gambia-Nigeria inclusive) with the emphasis upon species of ecological or economic importance.

HUMAN GEOGRAPHY

Attah-Poku, Agyemang. *African Ethnicity: History, Conflict Management, Resolution, and Prevention*. Lanham, Md.: University Press of America, 1997. Has chapters with titles such as "Ethnic History and Composition," and "Role of Ethnicity in the Past." It also discusses a number of instances of ethnic skirmishes and conflicts in Africa.

Beckwith, Carol, Angela Fisher, and Graham Hancock. *African Ark: People and Ancient Cultures of Ethiopia and the Horn of Africa*. New York: Harry N. Abrams, 1990. Photographers Beckwith and Fisher captured the exotic and natural beauty of the people of Ethiopia and the surrounding area. Hancock covers the history back to the early pre-Christian era.

Haskins, James, Jim Haskins, and Joann Biondi. *From Afar to Zulu: A Dictionary of African Cultures*. New York: Walker, 1998. Introduction to African cultures, describing the history, traditions, social structure, and daily life of some thirty ethnic groups. It also lists languages, populations, primary foods, and includes a section on Africa's lost cultures.

Legum, Colin. *Africa Since Independence*. Bloomington: Indiana University Press, 1999. Assesses Africa's experience since independence and offers predictions about the continent's future. It examines Africa's struggle for democracy, mounting economic problems, and AIDS.

Roberts, David. "Mali's Dogon People." *National Geographic* 178, no. 4 (October, 1990): 100-127. An excellent pictorial essay on the cliff-dwelling Dogon of eastern Mali, who are world-renowned for their wooden carvings and dance masks.

Stone, Martin. *The Agony of Algeria*. New York: Columbia University Press, 1997. Examines Algeria's history, from the founding of the Berber kingdoms, 130 years of French rule, and the devastating war for independence gained in 1962 to the present. It makes intelligible the current crisis tearing at the fabric of the country's society, while offering an analysis of the social, economic, and political challenges ahead.

ECONOMIC GEOGRAPHY

Byerlee, Derek. *Africa's Emerging Maize Revolution*. Boulder, Colo.: Lynne Rienner, 1997. Includes chapters on such topics as Africa's food crisis, Zimbabwe's emerging maize revolution, maize technology and productivity in Malawi, and maize productivity in Nigeria.

Fratkin, Elliot. *Ariaal Pastoralists of Kenya: Surviving Drought and Development in Africa's Arid Lands*. Needham Heights, Mass.: Allyn and Bacon, 1997. Presents the story of how one society of livestock herders in northern Kenya have adapted to and survived both natural and human-induced disasters of recent times, including drought and famine.

Mortimore, Michael. *Roots in the African Dust: Sustaining the Sub-Saharan Drylands.* New York: Cambridge University Press, 1998. Based on studies from East and West Africa, this study rejects the notion of runaway desertification driven by population growth and inappropriate land use and proposes solutions.

REGIONAL GEOGRAPHY

Zich, Arthur. "Modern Botswana, The Adopted Land." *National Geographic* 178, no. 6 (December, 1990): 70-97. An account of conditions in Botswana, including its democracy, current economy, its trade in diamonds and beef, and the Okavango Delta and its wildlife.

Zwingle, Erla. "Morocco." *National Geographic* 190, no. 4 (October, 1996): 98-125. Overview of modern Morocco, its history, its current economic conditions, and, in particular, its Berber people and their culture.

EUROPE

PHYSICAL GEOGRAPHY

Blake, S. F., and Alice C. Atwood. *Geographical Guide to Floras of the World: Western Europe.* Port Jervis, N.Y.: Lubrecht and Cramer, 1974. Extensive guide to the floras of areas such as Scandinavia, the Low Countries, the British Isles, Iberia, France, Italy, and the Netherlands.

Kuusia, K. *Forest Resources in Europe.* New York: Cambridge University Press, 1995. Provides a detailed country-by-country account of the increase in forest resources in Europe over the past forty years and what needs to be done to preserve the sustainability and biodiversity of Europe's forest ecosystems.

HUMAN GEOGRAPHY

Germek, Bronislaw. *The Common Roots of Europe.* Translated by S. Mitchell and R. Hunt. Oxford, England: Polity Press, 1997. Discusses unity, variety, and collective identity in medieval Europe, social and economic structures in East and West, and the continuity and change in European identity in the intervening centuries.

Haudry, Jean. *The Indo-Europeans.* Washington, D.C.: Scott-Townsend, 1998. Study of the roots of the Indo-European peoples emphasizing Europe, their migrations, and evolution into the present day.

Kiernan, Victor. *Lords of Human Kind: European Attitudes to Other Cultures in the Imperial Age.* London: Serif and Pixel Press, 1996. Using a great array of sources—missionaries' memoirs, the letters of diplomats' wives, explorers' diaries, and the work of writers as diverse as Voltaire, William

Makepeace Thackeray, Oliver Goldsmith and Rudyard Kipling—the author searches the full range of European attitudes to other peoples.

Unwin, Tim. *A European Geography.* Reading, Mass.: Addison-Wesley, 1998. Chapters in this volume cover a wide swath of Europe such as the peopling of Europe, the Celts, the peopling of Finland, European languages, religious dimensions of Europeans, and cultural landscapes.

ECONOMIC GEOGRAPHY

Grant, Wyn. *The Common Agricultural Policy.* New York: St. Martin's Press, 1997. Examines the European Common Agricultural Policy and its impact on trade between the United States and Europe. This study argues for a new set of objectives designed to deliver effective agricultural production at an acceptable cost and attuned to the growing concerns of citizens about food quality.

Holden, Mike J., and David Garrod. *The Common Fisheries Policy.* Williston, Vt.: Blackwell, 1996. Focuses on the conservation policy because it generated the most controversy, which continues to intensify even as fish stocks deteriorate. For many the conservation policy is the Common Fisheries Policy, apparently a disastrous failure.

Laux, James Michael. *The European Automobile Industry.* Indianapolis, Ind.: Macmillan, 1992. Looks at motor vehicle manufacturing on the Continent from 1890 to the present, paying particular attention to the postwar spurt of growth that established which of Europe's various automakers would prevail. He examines how European factory owners emulated American success in production and sales between the wars, how the postwar market boom chipped away at American dominance of the industry, and how Japanese models in turn began to cut into the world market in the 1980's.

REGIONAL GEOGRAPHY

Belt, Don. "Sweden." *National Geographic* 184, no. 2 (August, 1993): 2-35. Overview of modern Sweden, its culture and how it relates to the world, especially its closest neighbors, Denmark, Finland, and Norway.

Coniff, Richard. "Ireland." *National Geographic* 186, no. 3 (September, 1994): 2-36. An overview of present-day Ireland. Economic conditions have improved since manufacturing surpassed farming in the island nation.

Keillor, Garrison. "Civilized Denmark." *National Geographic* 194, no. 1 (July, 1998): 50-73. An overview of present-day Denmark and its society.

Vulliamy, Ed. "Romania's New Day." *National Geographic* 194, no. 3 (September, 1998): 35-59. An overview of Romania and its postcommunist society and the changes that are occurring.

Ward, Andrew. "Scotland." *National Geographic* 190, no. 3 (September, 1996): 2-27. Overview of modern Scotland, its history, and its current yearning for independence from England and the United Kingdom.

ASIA

PHYSICAL GEOGRAPHY

Hornocker, Maurice. "Siberian Tigers." *National Geographic* 191, no. 2 (February, 1997): 100-109. Only a few hundred survive in the wild. While zoos work to maintain the animal's genetic diversity, Russian and American scientists are pooling their efforts in the fight to save this magnificent creature from extinction. This article outlines the struggle, which includes the effects of poaching and habitat destruction.

Hutchison, Charles S. *Southeast Asian Oil, Gas, Coal and Mineral Deposits.* New York: Oxford University Press, 1996. Includes chapters on topics such as the oil and gas basins of Malaysia, Indonesia, and the Philippines, and coal, iron ore, tungsten, and tin deposits.

Knott, Cheryl. "Orangutans." *National Geographic* 194, no. 2 (August, 1998): 30-57. A study of a family of orangutans in Gunung Palung National Park near the west coast of Borneo.

Laidler, Liz, and Keith Laidler. *China's Threatened Wildlife.* Poole, Dorset, England: Blandford Press, 1999. This profiles twenty of China's more attention-getting endangered species: sixteen mammals, two birds, the giant salamander, and the Chinese alligator. It opens with a chapter describing China's eight distinct vegetation zones, ranging from tropical rainforest to alpine.

Moullade, Michel, and A. E. M. Naim. *Phanerozoic Geology of the World.* New York: Elsevier Science, 1991. Has chapters with titles such as "Southern Africa," "India," "Pakistan," "Late Precambrian and Paleozoic Rocks of Iran and Afghanistan," and "China."

Pant, Govind B., and Rupa K. Kumar. *Climates of South Asia.* New York: John Wiley and Sons, 1997. Explores the climates of countries in Southern Asia—India, Pakistan, Sri Lanka, Bangladesh, Nepal, Bhutan, and a few island countries of the Indian Ocean—using charts, diagrams, and data.

Schaller, George B. *Wildlife of the Tibetan Steppe.* Chicago: University of Chicago Press, 1998. Provides a detailed look at the flora and fauna of the Chang Tang, a remote Tibetan steppe. The plains ungulates are the main focus, especially the Tibetan antelope.

Verma, R. K. *Geodynamics of the Indian Peninsula and the Indian Margin.* Rotterdam, Netherlands: A. A. Balkema, 1991. On the geological history and evolution of the Indian Continental Shelf. Gravity fields, geology, and tectonics, radioactivity and heat sources, seismicity, and geodynamics of the Himalayas.

Ward, Geoffrey C. "Making Room for Wild Tigers." *National Geographic* 192, no. 6 (December, 1997): 2-35. An analysis of the work being done to accommodate all five subspecies of tigers in the increasingly densely populated areas of Asia where the animal is found.

_____. "India's Wildlife Dilemma." *National Geographic* 181, no. 5 (May, 1992): 2-29. The key problem is that growing numbers of poverty-stricken farmers compete for land with diverse wildlife species. This is threatening the future of India's unique natural heritage.

Wenshi, Pan. "New Hope for China's Giant Pandas." *National Geographic* 187, no. 2 (February, 1995): 100-115. Out of perhaps 1,200 pandas that remain in China, about 230 live in the Qin Ling area in Shaanxi Province in central China at elevations between 4,000 feet (1,200 meters) and 10,000 feet (3,000 meters). This is a look at a small family of pandas in that area.

HUMAN GEOGRAPHY

Kublin, Michael, and Hyman Kublin. *India.* Boston: Houghton Mifflin, 1991. Introduces the history and civilization of India. It includes a discussion of the problems facing Pakistan and Bangladesh.

Lardy, Nicholas R. *Agriculture in China's Modern Economic Development.* New York: Cambridge University Press, 1984. Explores the relationship between the Chinese peasantry, who are the fundamental base of support for the revolutionary Chinese Communist Party, and the state-led economic system established by the Party after 1949.

Schirokauer, Conrad. *A Brief History of Chinese Civilization.* Orlando, Fla.: HBJ College & School Division, 1991. Includes considerable material on the classical civilization of China, including Confucius, the Buddhist period, and the peoples.

_____. *A Brief History of Japanese Civilization.* Orlando, Fla.: HBJ College & School Division, 1993. Includes discussion of Shinto, samurai, the aristocracy, and even the Mongol invasion.

Songoiao, Zhao. *Geography of China: Environment, Resources, Population, and Development.* New York: John Wiley and Sons, 1994. Using a systematic and regional approach, this volume offers a comprehensive depiction of official population numbers, land and resource usage in the face of sobering population increase, population problems including ethnic structure and family planning, and a pattern of historical and economic development over China's long and interesting history.

ECONOMIC GEOGRAPHY

Gamaut, Rose Gregory, Guo Shutian, and Ma Guonon, eds. *The Third Revolution in the Chinese Countryside.* New York: Cambridge University Press, 1995. First section covers the issues of poverty in China and feeding the population. The second section describes the agricultural markets in China and the price reform of agricultural products. The next two parts discuss international and regional issues of China's agricultural economy.

Kalirajan, Kail P., ed. *Productivity and Growth in Chinese Agriculture.* New York: St. Martin's Press, 1999. Gauges the impact of economic and institutional reforms on agricultural productivity in China using the most recent farm household survey data. Results demonstrate the dynamic nature of Chinese farm households, particularly in relation to the changing demands placed on agriculture, especially the grain sector.

Pecht, Michael G., Wang Yong Wen, and Jiang Jun Lu. *The Electronics Industry in China.* Boca Raton, Fla.: CRC Press, 1999. Documents the technologies, capabilities, and infrastructure that has made China a major player in the Asian electronics industry.

Van Der Eng, Pierre. *Agricultural Growth in Indonesia Since 1880: Productivity Change and Policy Impact Since 1880.* New York: St. Martin's Press, 1996. Assesses long-term trends in agricultural production and productivity in Indonesia since 1880, providing an inventory of agricultural policies. It evaluates the impact of these policies on agricultural production, especially production of the country's main food and export crops. Appendices with statistics on prices, employment, livestock, and arable land.

REGIONAL GEOGRAPHY

Allen, Thomas B. "Turkey." *National Geographic* 185, no. 5 (May, 1994): 2-35. Overview of modern Turkey including discussion of its history, roots of the modern-day state, the legacy of Kemal Ataturk, and the tension between Islamic Turkey, which exists in the countryside, and secular, urban Turkey. There is some discussion of the Kurdish minority.

Cockburn, Andrew. "Yemen." *National Geographic* 197, no. 4 (April, 2000): 30-53. An overview of modern Yemen, the land of the Queen of Sheba. It shows that Yemen is a land where a very traditional Arab culture is still dominant.

McCarry, John. "The Promise of Pakistan." *National Geographic* 192, no. 4 (October, 1997): 49-73. An overview of modern Pakistan, its roots, cultures, peoples, geography, agriculture, and its problems.

Reid, T. R. "Malaysia." *National Geographic* 192, no. 2 (August, 1997): 100-121. Overview of Malaysia, a mix of Muslim Malays, Buddhist Chinese, and Hindus, and its more recent development.

Theroux, Peter. "Syria, Behind the Mask." *National Geographic* 190, no. 1 (July, 1996): 106-131. An overview of modern Syria. It tends to focus on the mellowing of the current regime and its reaching out to the West.

Vesilind, Prit J. "Sri Lanka." *National Geographic* 191, no. 1 (January, 1997): 111-133. Overview of modern Sri Lanka including the strife which exists between Hindu Tamils and the Buddhist Sinhalese.

Ward, Geoffrey C. "India." *National Geographic* 191, no. 5 (May, 1997): 2-57. Overview of modern India, its complexity, diverse peoples, large population, its great poverty, and its many accomplishments.

Waterlow, Julia. *China.* New York: Bookwright Press, 1990. Introduction to the geography, climate, schools, sports, food, recreation, and culture of China.

AUSTRALIA, PACIFIC, AND ANTARCTICA

PHYSICAL GEOGRAPHY
Blainey, Geoffrey. *Rush That Never Ended: A History of Australian Mining.* 4th ed. Melbourne, Australia: Melbourne University Press, 1993. Australia is one of the world's great sources of mineral treasure. The finding and development of minerals, oil, and natural gas have influenced Australian racial attitudes, unionism, religious life, law, and politics.

Conacher, Jeannette, and Arthur Conacher, eds. *Rural Land Degradation in Australia.* New York: Oxford University Press, 1995. Examines the degradation of Australia's ecosystems, the problems associated with the increasing use of synthetic chemicals, and the direct and underlying causes of land degradation. It also looks at broader social and economic implications, and places the nature of the overall problem in its global context.

Darcavel, John. *Fashioning Australia's Forests.* New York: Oxford University Press, 1996. Weaves together the story of industrial development and forest use with the slow acceptance of the case for forest conservancy.

Flannery, Tim F. *Mammals of the Southwest Pacific and Moluccan Islands.* Ithaca, N.Y.: Cornell University Press, 1995. Draws together the results of his five-year field survey and literature review on the mammals of an area extending from the islands just east of Sulawesi (Celebes, Indonesia) in the Moluccas, to the Cook Islands in the central South Pacific, north to Micronesia, and south to New Zealand, but excluding New Guinea.

Hodgson, Bryan. "Antarctica: A Land of Isolation No More." *National Geographic* 177, no. 4 (April, 1990): 2-51. Examination of the scientific research there and the controversies revolving around tourism, mineral exploitation, and water and atmospheric pollution.

Kanze, Edward. *Kangaroo Dreaming: An Australian Wildlife Odyssey.* New York: Random House, 2000. Detailed look, in the form of a travelogue, at the fauna of Australia.

Mueller-Dombois, Dieter, and F. Raymond Fosberg. *Vegetation of the Tropical Pacific Islands.* New York: Springer-Verlag, 1998. Extensive survey of the vegetation of the Pacific Islands, including the island of New Guinea, with illustrations.

Smith, David. *Water in Australia: Resources and Management.* New York: Oxford University Press, 1999. Outlines the nature of the resource, past management practices, policy, and the outlook for the future.

Soper, Tony. *Antarctica: A Guide to the Wildlife.* Old Saybrook, Conn.: Globe Pequot Press, 1997. The storm-tossed Southern Ocean and the inhospitable landscape of Antarctica combine to form one of the last true wildernesses on Earth. They are also home to vast numbers of animals, from the tiny shrimp of the zooplankton to the penguins, albatrosses, seals, and great whales for which this region is famed.

HUMAN GEOGRAPHY

Belich, James. *Making Peoples: A History of the New Zealanders: From Polyne-sian Settlement to the End of the Nineteenth Century.* Honolulu: University of Hawaii Press, 1997. Account of the active and dynamic Maori engagement with the history of New Zealand both before and after British settlement.

Darien-Smith, Kate, and David Lowe. *The Australian Outback and Its People.* Orlando, Fla.: Raintree Steck-Vaughn, 1995. The large, dry regions of Australia, known as the outback, are introduced through brief, slight discussions of their history, environment, inhabitants, and future. The aboriginal culture and the European impact on it are explored at greater length.

Lindstrom, Lamont, and Geoffrey M. White. *Culture, Custom and Tradi-tion: Cultural Policy in Melanesia.* Suva, Fiji: Institute of Pacific Studies, 1994. Looks broadly at cultural development programs and policies in three Melanesian countries: Papua New Guinea, Solomon Islands, and Vanuatu. With more than a thousand distinct linguistic-cultural groups, Melanesia is the most culturally diverse area in the world. Local and national attempts to protect and promote this rich concentration of cultural traditions have produced some novel experiments in cultural development.

New Politics in the South Pacific. Suva, Fiji: Institute of Pacific Studies, 1994. Written almost entirely by Pacific Islanders, many of whom are active in the political process, this volume examines the evolving impact of women in politics, of electronic media, of sovereignty movements on one hand and federation movements on the other. It also examines the search for forms of political and constitutional association be-tween small countries and large metropolitan powers that yield both the dignity of independence and the security and diversity of belong-ing to large systems.

Nile, Richard, and Christian Clerk. *Australia, New Zealand, and the South Pacific.* New York: Facts on File, 1996. Taking migration as one of its themes, this Atlas traces the great movements of people into this re-gion from earliest times. It describes the complex societies and cul-tures that evolved in the Pacific and explores the cultural differences between the three major cultural areas, Melanesia, Micronesia, and Polynesia. It also examines the founding myths that shaped Australia and New Zealand's emergent national identities and looks at the great changes that have taken place since 1945.

ECONOMIC GEOGRAPHY

King, Michael G. *Fisheries in the Economy of the South Pacific.* Suva, Fiji: Insti-tute of Pacific Studies, 1991. Describes resources, methods, and man-agement of fisheries in the South Pacific.

May, Dawn. *Aboriginal Labour and the Cattle Industry: Queensland from White*

Settlement to the Present. New York: Cambridge University Press, 1994. Uncovers the central role of Aboriginal labor in the Queensland cattle industry from first contact to the present. It shows that the use of Aboriginal labor was a complex process involving a high degree of state intervention.

REGIONAL GEOGRAPHY

McKnight, Tom L. *Oceania: The Geography of Australia, New Zealand and the Pacific Islands.* Upper Saddle River, N.J.: Prentice Hall, 1998. Introduces the geography of the Pacific region in broad terms, then focuses on Australia, blending in discussion of the industries, population, contemporary issues, and problems as they relate to geography. New Zealand's land, people, and regions are discussed next, and the smaller islands of the Pacific receive one chapter's discussion.

Dana P. McDermott

APPENDICES

1. Regions of the World

(Numbers are keyed to regional maps that follow.)

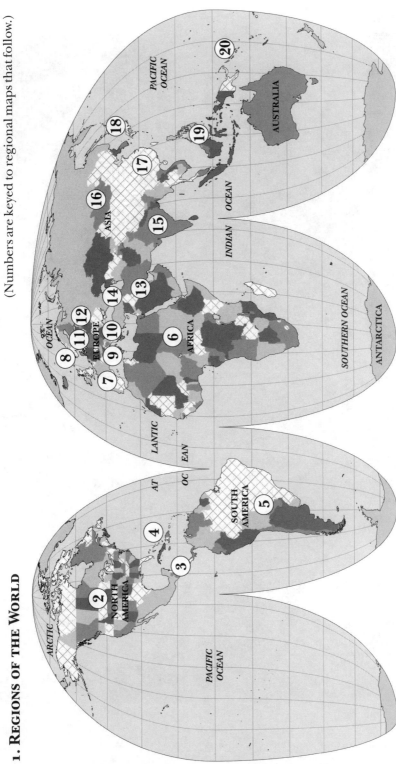

2. NORTH AMERICA

Alaska

Yukon Territory

Northwest Territories

Nunavut

British Columbia

Alberta

CANADA

Saskatchewan

Manitoba

Ontario

Quebec

Newfoundland

Prince Edward Island

New Brunswick

Washington

Montana

North Dakota

Minnesota

Ottawa*

Maine

Nova Scotia

Oregon

Idaho

Michigan

Wisconsin

New Hampshire

Vermont

Massachusetts

California

Nevada

Wyoming

South Dakota

New York

Rhode Island

Connecticut

Utah

Nebraska

Iowa

Pennsylvania

New Jersey

UNITED STATES

Colorado

Illinois

Indiana

Ohio

West Virginia

Washington D.C.

Delaware

Arizona

New Mexico

Kansas

Missouri

Kentucky

Virginia

Maryland

Oklahoma

Tennessee

North Carolina

Arkansas

South Carolina

Baja California

Sonora

Texas

Mississippi

Alabama

Georgia

Atlantic Ocean

Pacific Ocean

Chihuahua

Louisiana

Baja California Sur

Coahuila

Florida

MEXICO

Sinaloa

Durango

Nuevo Leon

Gulf of Mexico

Zacatecas

Tamaulipas

Nayarit

San Luis Potosi

Hawaii

Jalisco

Colima

Michoacan

Mexico City

Yucatán

Campeche

Quintana Roo

Guerrero

Veracruz

Tabasco

Oaxaca

Chiapas

Caribbean Sea

Key to States
(shown by numbers on map)

1 Aguascalientes
2 Guanajuato
3 Querétaro
4 Hidalgo
5 Mexico
6 Tlaxcala
7 Distrito Federal
8 Morelos
9 Puebla

3. Central America

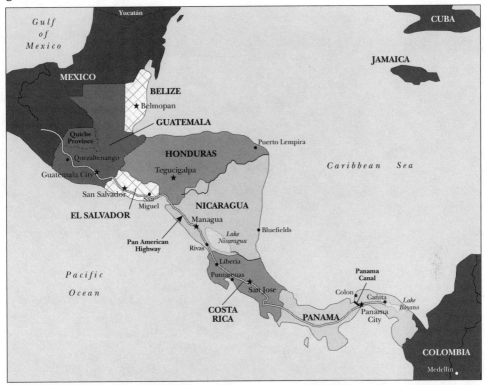

4. Caribbean

5. SOUTH AMERICA

PANAMA

VENEZUELA

GUYANA

SURINAME

FRENCH
GUIANA

North

Atlantic

Ocean

Medellin

Bogotá ★

Cali

COLOMBIA

*Galápagos
Islands*

Quito ★

ECUADOR

A m a z o n   Basin

Belem

PERU

BRAZIL

Lima ★

BOLIVIA

La Paz ★

Arica

Brasilia ★

*South
Pacific
Ocean*

PARAGUAY

Sao Paulo ●   Rio de Janeiro ●

Asuncion ★

CHILE

A N D E S   M O U N T A I N S

ARGENTINA

URUGUAY

Santiago ★

Buenos
Aires ★

Montevideo ★

*South
Atlantic
Ocean*

Falkland Islands

6. AFRICA

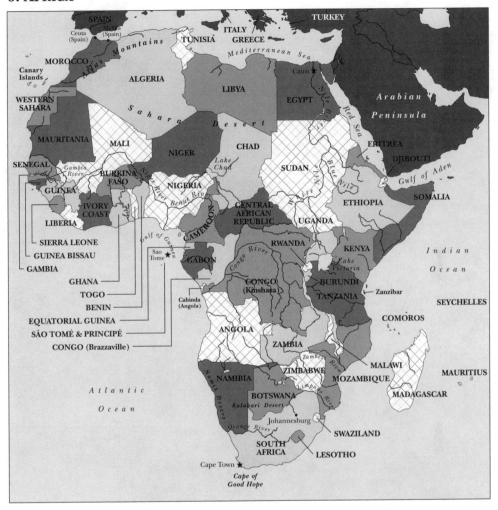

576

7. WESTERN EUROPE

Micro States
1. Andorra
2. Liechtenstein
3. Monaco
4. San Marino
5. Vatican City

8. SCANDINAVIA

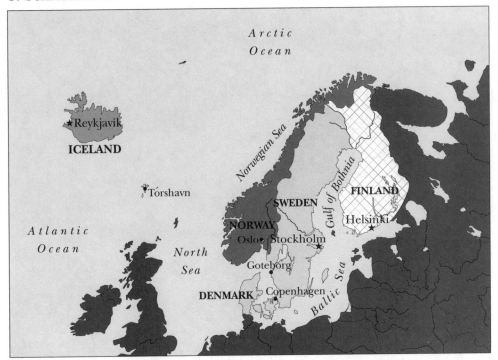

9. MEDITERRANEAN EUROPE

10. Balkan Nations

11. CENTRAL EUROPE

12. Former Soviet European Nations

13. MIDDLE EAST

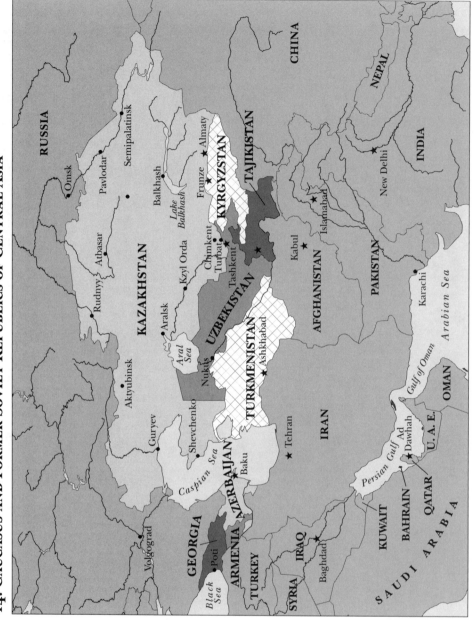

14. Caucasus and Former Soviet Republics of Central Asia

15. SOUTH ASIA

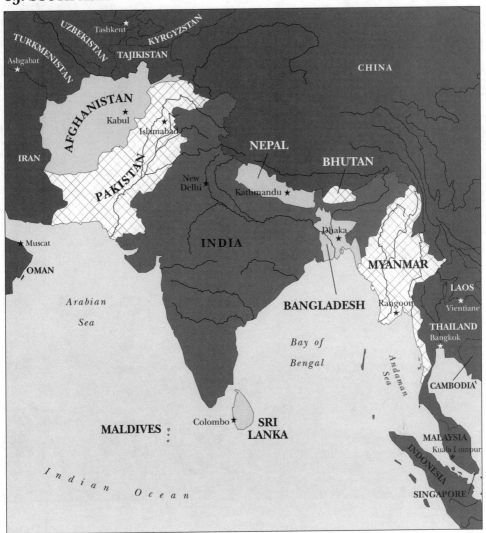

16. Mongolia and Asian Russia

17. EAST ASIA

18. JAPAN

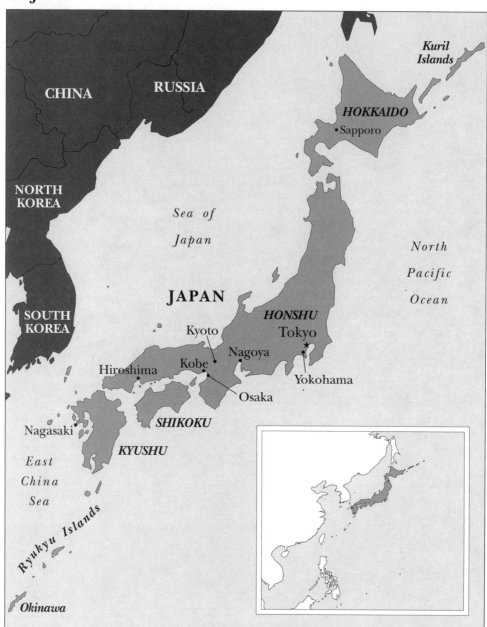

19. SOUTHEAST ASIA

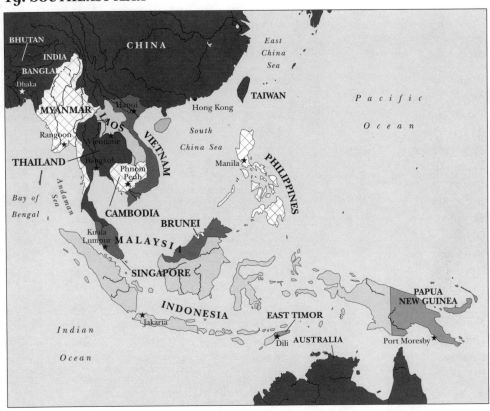

20. SOUTH PACIFIC AND AUSTRALASIA

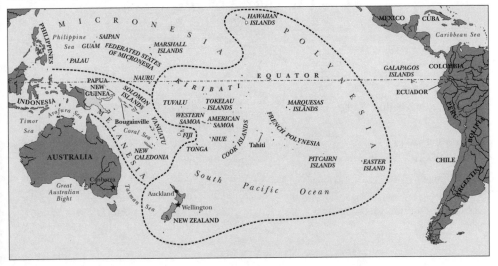

THE WORLD'S OCEANS AND SEAS

Name	Approximate Area		Average Depth	
	Sq. Miles	Sq. Km.	Feet	Meters
Pacific Ocean	64,000,000	165,760,000	13,215	4,028
Atlantic Ocean	31,815,000	82,400,000	12,880	3,926
Indian Ocean	25,300,000	65,526,700	13,002	3,963
Arctic Ocean	5,440,200	14,090,000	3,953	1,205
Mediterranean and Black Seas	1,145,100	2,965,800	4,688	1,429
Caribbean Sea	1,049,500	2,718,200	8,685	2,647
South China Sea	895,400	2,319,000	5,419	1,652
Bering Sea	884,900	2,291,900	5,075	1,547
Gulf of Mexico	615,000	1,592,800	4,874	1,486
Okhotsk Sea	613,800	1,589,700	2,749	838
East China Sea	482,300	1,249,200	617	188
Hudson Bay	475,800	1,232,300	420	128
Japan Sea	389,100	1,007,800	4,429	1,350
Andaman Sea	308,100	797,700	2,854	870
North Sea	222,100	575,200	308	94
Red Sea	169,100	438,000	1,611	491
Baltic Sea	163,000	422,200	180	55

MAJOR LAND AREAS OF THE WORLD

Area	Approximate Land Area		Portion of World Total
	Sq. Mi.	Sq. Km.	
World	57,308,738	148,429,000	100.0%
Asia (including Middle East)	17,212,041	44,579,000	30.0
Africa	11,608,156	30,065,000	20.3
North America	9,365,290	24,256,000	16.3
Central America, South America, and Caribbean	6,879,952	17,819,000	8.9
Antarctica	5,100,021	13,209,000	8.9
Europe	3,837,082	9,938,000	6.7
Oceania, including Australia	2,967,966	7,687,000	5.2

Major Islands of the World

		Area	
Island	*Location*	*Sq. Mi.*	*Sq. Km*
Greenland	North Atlantic Ocean	839,999	2,175,597
New Guinea	Western Pacific Ocean	316,615	820,033
Borneo	Western Pacific Ocean	286,914	743,107
Madagascar	Western Indian Ocean	226,657	587,042
Baffin	Canada, North Atlantic Ocean	183,810	476,068
Sumatra	Indonesia, northeast Indian Ocean	182,859	473,605
Hōnshū	Japan, western Pacific Ocean	88,925	230,316
Great Britain	North Atlantic Ocean	88,758	229,883
Ellesmere	Canada, Arctic Ocean	82,119	212,688
Victoria	Canada, Arctic Ocean	81,930	212,199
Sulawesi (Celebes)	Indonesia, western Pacific Ocean	72,986	189,034
South Island	New Zealand, South Pacific Ocean	58,093	150,461
Java	Indonesia, Indian Ocean	48,990	126,884
North Island	New Zealand, South Pacific Ocean	44,281	114,688
Cuba	Caribbean Sea	44,218	114,525
Newfoundland	Canada, North Atlantic Ocean	42,734	110,681
Luzon	Philippines, western Pacific Ocean	40,420	104,688
Iceland	North Atlantic Ocean	39,768	102,999
Mindanao	Philippines, western Pacific Ocean	36,537	94,631
Ireland	North Atlantic Ocean	32,597	84,426
Hokkaido	Japan, western Pacific Ocean	30,372	78,663
Hispaniola	Caribbean Sea	29,355	76,029
Tasmania	Australia, South Pacific Ocean	26,215	67,897
Sri Lanka	Indian Ocean	25,332	65,610
Sakhalin (Karafuto)	Russia, western Pacific Ocean	24,560	63,610
Banks	Canada, Arctic Ocean	23,230	60,166
Devon	Canada, Arctic Ocean	20,861	54,030
Tierra del Fuego	Southern tip of South America	18,605	48,187
Kyūshū	Japan, western Pacific Ocean	16,223	42,018
Melville	Canada, Arctic Ocean	16,141	41,805
Axel Heiberg	Canada, Arctic Ocean	15,779	40,868
Southampton	Hudson Bay, Canada	15,700	40,663

COUNTRIES OF THE WORLD

Country	Map	Region	Population	Area Square Miles	Area Square Kilometers
Afghanistan	15	Asia	28,717,213	249,935	647,500
Albania	10	Europe	3,582,205	11,098	28,750
Algeria	6	Africa	32,818,500	919,352	2,381,740
Andorra	7	Europe	69,150	174	450
Angola	6	Africa	10,766,471	481,226	1,246,700
Antigua and Barbuda	4	Caribbean	67,897	170	440
Argentina	5	South America	38,740,807	1,068,020	2,766,890
Armenia	14	Europe	3,326,448	11,503	29,800
Australia	20	Australia	19,731,984	2,967,124	7,686,850
Austria	7	Europe	8,188,207	32,369	83,858
Azerbaijan	14	Asia	7,830,764	33,428	86,600
Bahamas	4	Caribbean	297,477	5,381	13,940
Bahrain	13	Asia	667,238	239	620
Bangladesh	15	Asia	138,448,210	55,584	144,000
Barbados	4	Caribbean	277,264	166	430
Belarus	12	Europe	10,322,151	80,134	207,600
Belgium	7	Europe	10,289,088	11,777	30,510
Belize	3	Central America	266,440	8,863	22,960
Benin	6	Africa	7,041,490	43,471	112,620
Bhutan	15	Asia	2,139,549	18,142	47,000
Bolivia	5	South America	8,586,443	424,052	1,098,580
Bosnia and Herzegovina	10	Europe	3,989,018	19,776	51,233
Botswana	6	Africa	1,573,267	231,743	600,370
Brazil	5	South America	182,032,604	3,285,618	8,511,965
Brunei	13	Asia	358,098	2,227	5,770
Bulgaria	10	Europe	7,537,929	42,811	110,910
Burkina Faso	6	Africa	13,228,460	105,841	274,200
Burundi	6	Africa	6,096,156	10,742	27,830

Country	Map	Region	Population	Area Square Miles	Area Square Kilometers
Cambodia	19	Asia	13,124,764	69,881	181,040
Cameroon	6	Africa	15,746,179	183,520	475,440
Canada	2	North America	32,207,113	3,850,790	9,976,140
Cape Verde	6	Africa	412,137	1,556	4,030
Central African Republic	6	Africa	3,683,538	240,470	622,980
Chad	6	Africa	9,253,493	495,624	1,284,000
Chile	5	South America	15,665,216	292,183	756,950
China, People's Republic of	17	Asia	1,286,975,468	3,704,427	9,596,960
Colombia	5	South America	41,662,073	439,619	1,138,910
Comoros	6	Africa	632,948	838	2,170
Congo (Brazzaville)	6	Africa	2,954,258	132,012	342,000
Congo (Kinshasa)	6	Africa	56,625,039	905,328	2,345,410
Costa Rica	3	Central America	3,896,092	19,725	51,100
Côte d'Ivoire	6	Africa	16,962,491	124,470	322,460
Croatia	10	Europe	4,422,248	21,824	56,538
Cuba	4	Caribbean	11,263,429	42,792	110,860
Cyprus	9	Europe	771,657	3,571	9,250
Czech Republic	11	Europe	10,249,216	30,379	78,703
Denmark	8	Europe	5,384,384	16,634	43,094
Djibouti	6	Africa	457,130	8,492	22,000
Dominica	4	Caribbean	69,655	290	750
Dominican Republic	4	Caribbean	8,715,602	18,810	48,730
East Timor	19	Asia	997,853	7,336	19,000
Ecuador	5	South America	13,710,234	109,454	283,560
Egypt	6	Africa	74,718,797	386,560	1,001,450
El Salvador	3	Central America	6,470,379	8,121	21,040
Equatorial Guinea	6	Africa	510,473	10,827	28,050
Eritrea	6	Africa	4,362,254	46,830	121,320
Estonia	12	Europe	1,408,556	17,457	45,226

(continued)

COUNTRIES OF THE WORLD — *continued*

Country	Map	Region	Population	Square Miles	Square Kilometers
Ethiopia	6	Africa	66,557,553	435,071	1,127,127
Fiji	20	Pacific Islands	868,531	7,052	18,270
Finland	8	Europe	5,190,785	130,094	337,030
France	7	Europe	60,180,529	211,154	547,030
Gabon	6	Africa	1,321,560	103,321	267,670
Gambia	6	Africa	1,501,050	4,362	11,300
Georgia	14	Europe	4,934,413	26,904	69,700
Germany	7	Europe	82,398,326	137,767	356,910
Ghana	6	Africa	20,467,747	92,076	238,540
Greece	9	Europe	10,665,989	50,929	131,940
Grenada	4	Caribbean	89,258	131	340
Guam	20	Pacific Islands	163,941	212	549
Guatemala	3	Central America	13,909,384	42,032	108,890
Guinea	6	Africa	9,030,220	94,902	245,860
Guinea-Bissau	6	Africa	1,360,827	13,942	36,120
Guyana	5	South America	702,100	82,978	214,970
Haiti	4	Caribbean	7,527,817	10,712	27,750
Honduras	3	Central America	6,669,789	43,267	112,090
Hungary	11	Europe	10,045,407	35,910	93,030
Iceland	8	Europe	280,798	39,758	103,000
India	15	Asia	1,049,700,118	1,269,010	3,287,590
Indonesia	19	Asia	234,893,453	740,904	1,919,440
Iran	13	Asia	68,278,826	636,128	1,648,000
Iraq	13	Asia	24,683,313	168,710	437,072
Ireland	7	Europe	3,924,140	27,128	70,280
Israel	13	Asia	6,116,533	8,017	20,770
Italy	9	Europe	57,998,353	116,275	301,230
Jamaica	4	Caribbean	2,695,867	4,242	10,990
Japan	18	Asia	127,214,499	145,844	377,835
Jordan	13	Asia	5,460,265	34,436	89,213

Country	Map	Region	Population	Area Square Miles	Area Square Kilometers
Kazakhstan	14	Asia	16,763,795	1,048,878	2,717,300
Kenya	6	Africa	31,639,091	224,903	582,650
Kiribati	20	Pacific Islands	98,549	277	717
Korea, North	17	Asia	22,466,481	46,528	120,540
Korea, South	17	Asia	48,289,037	38,013	98,480
Kuwait	13	Asia	2,183,161	6,879	17,820
Kyrgyzstan	14	Asia	4,892,808	76,621	198,500
Laos	19	Asia	5,921,545	91,405	236,800
Latvia	12	Europe	2,348,784	24,743	64,100
Lebanon	13	Asia	3,727,703	4,014	10,400
Lesotho	6	Africa	1,861,959	11,715	30,350
Liberia	6	Africa	3,317,176	42,989	111,370
Libya	6	Africa	5,499,074	679,182	1,759,540
Liechtenstein	7	Europe	33,145	62	160
Lithuania	12	Europe	3,592,561	25,167	65,200
Luxembourg	7	Europe	454,157	998	2,586
Macedonia	10	Europe	2,063,122	9,779	25,333
Madagascar	6	Africa	16,979,744	226,597	587,040
Malawi	6	Africa	11,651,239	45,733	118,480
Malaysia	19	Asia	23,092,940	127,284	329,750
Maldives	15	Asia	329,684	116	300
Mali	6	Africa	11,626,219	478,640	1,240,000
Malta	9	Europe	400,420	124	320
Marshall Islands	20	Pacific Islands	56,429	70	181.3
Martinique	4	Caribbean	425,966	425	1,100
Mauritania	6	Africa	2,912,584	397,850	1,030,700
Mauritius	6	Africa	1,210,447	718	1,860
Mexico	2	North America	104,907,991	761,404	1,972,550
Micronesia	20	Pacific Islands	136,973	271	702
Moldova	12	Europe	4,439,502	13,008	33,700

(continued)

COUNTRIES OF THE WORLD — *continued*

Country	Map	Region	Population	Square Miles	Square Kilometers
Monaco	7	Europe	32,130	1	1.95
Mongolia	16, 17	Asia	2,712,315	604,090	1,565,000
Morocco	6	Africa	31,689,265	172,368	446,550
Mozambique	6	Africa	17,479,266	309,414	801,590
Myanmar (Burma)	15, 19	Asia	42,510,537	261,901	678,500
Namibia	6	Africa	1,927,447	318,611	825,418
Nauru	20	Pacific Islands	12,570	8	21
Nepal	15	Asia	26,469,569	54,349	140,800
Netherlands	7	Europe	16,150,511	16,029	41,526
New Zealand	20	Pacific Islands	3,951,307	103,710	268,680
Nicaragua	3	Central America	5,128,517	49,985	129,494
Niger	6	Africa	11,058,590	489,062	1,267,000
Nigeria	6	Africa	133,881,703	356,575	923,770
Norway	8	Europe	4,546,123	125,149	324,220
Oman	13	Asia	2,807,125	82,010	212,460
Pakistan	15	Asia	150,694,740	310,321	803,940
Palau	20	Pacific Islands	19,717	177	458
Panama	3	Central America	2,960,784	30,185	78,200
Papua New Guinea	20	Pacific Islands	5,295,816	178,212	461,690
Paraguay	5	South America	6,036,900	157,006	406,750
Peru	5	South America	28,409,897	496,095	1,285,220
Philippines	19	Asia	84,619,974	115,800	300,000
Poland	11	Europe	38,622,660	120,696	312,683
Portugal	7, 9	Europe	10,102,022	35,663	92,391
Qatar	13	Asia	817,052	4,415	11,437
Romania	11	Europe	22,271,839	91,675	237,500
Russia	12, 16	Europe/Asia	144,526,278	6,591,027	17,075,200
Rwanda	6	Africa	7,810,056	10,167	26,340
Saint Kitts and Nevis	4	Caribbean	38,763	104	269

Country	Map	Region	Population	Area Square Miles	Area Square Kilometers
Saint Lucia	4	Caribbean	162,157	239	620
Saint Vincent and Grenadines	4	Caribbean	116,812	131	340
Samoa	20	Pacific Islands	178,173	1,104	2,860
San Marino	7	Europe	28,119	23	60
São Tomé and Príncipe	6	Africa	175,883	371	960
Saudi Arabia	13	Asia	24,293,844	756,785	1,960,582
Senegal	6	Africa	10,580,307	75,729	196,190
Seychelles	6	Africa	80,469	176	455
Sierra Leone	6	Africa	5,732,681	27,692	71,740
Singapore	19	Asia	4,608,595	250	647.5
Slovakia	11	Europe	5,430,033	18,854	48,845
Slovenia	10	Europe	1,935,677	7,819	20,256
Solomon Islands	20	Pacific Islands	509,190	10,982	28,450
Somalia	6	Africa	8,025,190	246,137	637,660
South Africa	6	Africa	42,768,678	470,886	1,219,912
Spain	7, 9	Europe	40,217,413	194,834	504,750
Sri Lanka	15	Asia	19,742,439	25,325	65,610
Sudan	6	Africa	38,114,160	967,243	2,505,810
Suriname	5	South America	435,449	63,022	163,270
Swaziland	6	Africa	1,161,219	6,701	17,360
Sweden	8	Europe	8,878,085	173,686	449,964
Switzerland	7	Europe	7,318,638	15,938	41,290
Syria	13	Asia	17,585,540	71,479	185,180
Taiwan	17	Asia	22,603,000	13,888	35,980
Tajikistan	14	Asia	6,863,752	55,237	143,100
Tanzania	6	Africa	35,922,454	364,805	945,090
Thailand	19	Asia	64,265,276	198,404	514,000
Togo	6	Africa	5,429,299	21,921	56,790
Tonga	20	Pacific Islands	108,141	289	748

(continued)

COUNTRIES OF THE WORLD — *continued*

Country	Map	Region	Population	Square Miles	Square Kilometers
Trinidad and Tobago	4	Caribbean	1,104,209	1,980	5,130
Tunisia	6	Africa	9,924,742	63,153	163,610
Turkey	13	Europe/Asia	68,109,469	301,304	780,580
Turkmenistan	14	Asia	4,775,544	188,407	488,100
Tuvalu	20	Pacific Islands	11,305	10	26
Uganda	6	Africa	25,632,794	91,111	236,040
Ukraine	12	Europe	48,055,439	233,028	603,700
United Arab Emirates	13	Asia	2,484,818	31,992	82,880
United Kingdom	7	Europe	60,094,648	94,501	244,820
United States	2	North America	290,342,554	3,716,829	9,629,091
Uruguay	5	South America	3,413,329	68,021	176,220
Uzbekistan	14	Asia	25,981,647	172,696	447,400
Vanuatu	20	Pacific Islands	199,414	5,697	14,760
Vatican City	9	Europe	900	0.2	.44
Venezuela	5	South America	24,654,694	352,051	912,050
Vietnam	19	Asia	81,624,716	127,210	329,560
Western Sahara	6	Africa	261,794	102,676	266,000
Yemen	13	Asia	19,349,881	203,796	527,970
Yugoslavia	10	Europe	10,655,774	39,507	102,350
Zambia	6	Africa	10,307,333	290,507	752,610
Zimbabwe	6	Africa	12,576,742	150,764	390,580

Note: Population figures are October, 2002, estimates.
Source: U.S. Census Bureau, International Data Base.

THE WORLD'S LARGEST COUNTRIES BY AREA

			Area	
Rank	Country	Region	Sq. Miles	Sq. Km.
1	Russia	Europe/Asia	6,591,027	17,075,200
2	Canada	North America	3,850,790	9,976,140
3	United States	North America	3,716,829	9,629,091
4	China, People's Republic of	Asia	3,704,427	9,596,960
5	Brazil	South America	3,285,618	8,511,965
6	Australia	Australia	2,967,124	7,686,850
7	India	Asia	1,269,010	3,287,590
8	Argentina	South America	1,068,020	2,766,890
9	Kazakhstan	Asia	1,048,878	2,717,300
10	Sudan	Africa	967,243	2,505,810
11	Algeria	Africa	919,352	2,381,740
12	Congo (Kinshasa)	Africa	905,328	2,345,410
13	Mexico	North America	761,404	1,972,550
14	Saudi Arabia	Asia	756,785	1,960,582
15	Indonesia	Asia	740,904	1,919,440
16	Libya	Africa	679,182	1,759,540
17	Iran	Asia	636,128	1,648,000
18	Mongolia	Asia	604,090	1,565,000
19	Peru	South America	496,095	1,285,220
20	Chad	Africa	495,624	1,284,000
21	Niger	Africa	489,062	1,267,000
22	Angola	Africa	481,226	1,246,700
23	Mali	Africa	478,640	1,240,000
24	South Africa	Africa	470,886	1,219,912
25	Colombia	South America	439,619	1,138,910
26	Ethiopia	Africa	435,071	1,127,127
27	Bolivia	South America	424,052	1,098,580
28	Mauritania	Africa	397,850	1,030,700
29	Egypt	Africa	386,560	1,001,450
30	Tanzania	Africa	364,805	945,090

Source: U.S. Census Bureau, International Data Base.

THE WORLD'S SMALLEST COUNTRIES BY AREA

Rank	Country	Region	Sq. Miles	Sq. Km.
1	Vatican City*	Europe	0.2	.44
2	Monaco*	Europe	1	1.95
3	Nauru	Pacific Islands	8	21
4	Tuvalu	Pacific Islands	10	26
5	San Marino*	Europe	23	60
6	Liechtenstein*	Europe	62	160
7	Marshall Islands	Pacific Islands	70	181.3
8	Saint Kitts and Nevis	Caribbean	104	269
9	Maldives	Asia	116	300
10	Malta	Europe	124	320
11	Grenada	Caribbean	131	340
12	Saint Vincent and Grenadines	Caribbean	131	340
13	Barbados	Caribbean	166	430
14	Antigua and Barbuda	Caribbean	170	440
15	Andorra*	Europe	174	450
16	Seychelles	Africa	176	455
17	Palau	Pacific Islands	177	458
18	Guam	Pacific Islands	212	549
19	Bahrain	Asia	239	620
20	Saint Lucia	Caribbean	239	620
21	Singapore*	Asia	250	647.5
22	Micronesia	Pacific Islands	271	702
23	Kiribati	Pacific Islands	277	717
24	Tonga	Pacific Islands	289	748
25	Dominica	Caribbean	290	750
26	São Tomé and Príncipe	Africa	371	960
27	Martinique	Caribbean	425	1,100
28	Mauritius	Africa	718	1,860
29	Comoros	Africa	838	2,170
30	Luxembourg*	Europe	998	2,586

Note: Asterisks (*) denote countries on continents; all other countries are islands or island groups.
Source: U.S. Census Bureau, International Data Base.

THE WORLD'S LARGEST COUNTRIES BY POPULATION

Rank	Country	Region	Population
1	China	Asia	1,286,975,468
2	India	Asia	1,049,700,118
3	United States	North America	290,342,554
4	Indonesia	Asia	234,893,453
5	Brazil	South America	182,032,604
6	Pakistan	Asia	150,694,740
7	Russia	Europe/Asia	144,526,278
8	Bangladesh	Asia	138,448,210
9	Nigeria	Africa	133,881,703
10	Japan	Asia	127,214,499
11	Mexico	North America	104,907,991
12	Philippines	Asia	84,619,974
13	Germany	Europe	82,398,326
14	Vietnam	Asia	81,624,716
15	Egypt	Africa	74,718,797
16	Iran	Asia	68,278,826
17	Turkey	Europe/Asia	68,109,469
18	Ethiopia	Africa	66,557,553
19	Thailand	Asia	64,265,276
20	France	Europe	60,180,529
21	United Kingdom	Europe	60,094,648
22	Italy	Europe	57,998,353
23	Congo (Kinshasa)	Africa	56,625,039
24	Korea, South	Asia	48,289,037
25	Ukraine	Asia	48,055,439
26	South Africa	Africa	42,768,678
27	Mayanmar (Burma)	Asia	42,510,537
28	Colombia	South America	41,662,073
29	Spain	Europe	40,217,413
30	Argentina	South America	38,740,807

Source: U.S. Census Bureau, International Data Base. Updated October 10, 2002.

THE WORLD'S SMALLEST COUNTRIES BY POPULATION

Rank	Country	Region	Population
1	Vatican City	Europe	900
2	Tuvalu	Pacific Islands	11,305
3	Nauru	Pacific Islands	12,570
4	Palau	Pacific Islands	19,717
5	San Marino	Europe	28,119
6	Monaco	Europe	32,130
7	Liechtenstein	Europe	33,145
8	Saint Kitts and Nevis	Caribbean	38,763
9	Marshall Islands	Pacific Islands	56,429
10	Antigua and Barbuda	Caribbean	67,897
11	Andorra	Europe	69,150
12	Dominica	Caribbean	69,655
13	Seychelles	Africa	80,469
14	Grenada	Caribbean	89,258
15	Kiribati	Pacific Islands	98,549
16	Tonga	Pacific Islands	108,141
17	Saint Vincent and the Grenadines	Caribbean	116,812
18	Micronesia	Pacific Islands	136,973
19	Saint Lucia	Caribbean	162,157
20	Guam	Pacific Islands	163,941
21	São Tome and Principe	Africa	175,883
22	Samoa	Pacific Islands	178,173
23	Vanuatu	Pacific Islands	199,414
24	Western Sahara	Africa	261,794
25	Belize	Central America	266,440
26	Barbados	Caribbean	277,264
27	Iceland	Europe	280,798
28	Bahamas	Caribbean	297,477
29	Maldives	Asia	329,684
30	Brunei	Asia	358,098

Note: Population figures are October, 2002, estimates.
Source: U.S. Census Bureau, International Data Base.

THE WORLD'S MOST DENSELY POPULATED COUNTRIES

Rank	Country	Region	Persons per square Mile	Kilometer
1	Monaco	Europe	41,423.2	15,993.5
2	Singapore	Asia	18,481.7	7,135.8
3	Vatican City	Europe	5,698.0	2,200.0
4	Malta	Europe	3,207.2	1,238.3
5	Maldives	Asia	2,764.0	1,067.2
6	Bahrain	Asia	2,746.4	1,060.4
7	Bangladesh	Asia	2,579.6	996.0
8	Taiwan	Asia	1,810.2	698.9
9	Mauritius	Africa	1,681.2	649.1
10	Barbados	Caribbean	1,666.1	643.3

Note: Based on October, 2002, population estimates.
Source: U.S. Census Bureau, U.S. Department of Commerce.

THE WORLD'S LEAST DENSELY POPULATED COUNTRIES

Rank	Country	Region	Persons per square Mile	Kilometer
1	Mongolia	Asia	4.4	1.7
2	Namibia	Africa	5.7	2.2
3	Australia	Australasia	6.7	2.6
4	Suriname	South America	7.0	2.7
5	Botswana	Africa	7.0	2.7
6	Mauritania	Africa	7.0	2.7
7	Iceland	Europe	7.3	2.8
8	Libya	Africa	8.0	3.1
9	Guyana	South America	9.1	3.5
10	Canada	North America	9.1	3.5

Note: Based on October, 2002, population estimates.
Source: U.S. Census Bureau, U.S. Department of Commerce.

THE WORLD'S MOST POPULOUS CITIES

Rank	City	Country	Region	Population
1	Seoul	South Korea	East Asia	10,231,217
2	São Paulo	Brazil	South America	10,017,821
3	Mumbai (Bombay)	India	Asia	9,925,891
4	Jakarta	Indonesia	Asia	9,112,652
5	Moscow	Russia	Europe	8,368,449
6	Istanbul	Turkey	Europe	8,274,921
7	Mexico City	Mexico	North America	8,235,744
8	Shanghai	China	Asia	8,214,384
9	Tokyo	Japan	Asia	7,967,614
10	New York City	United States	North America	7,380,906
11	Beijing	China	Asia	7,362,426
12	Delhi	India	Asia	7,206,704
13	London	Great Britain	Europe	7,074,265
14	Cairo	Egypt	Africa	6,800,000
15	Teheran	Iran	Asia	6,750,043
16	Hong Kong	China	Asia	6,502,000
17	Bangkok	Thailand	Asia	5,882,000
18	Tianjin	China	Asia	5,855,044
19	Lima	Peru	South America	5,681,941
20	Rio de Janeiro	Brazil	South America	5,606,497
21	Bogotá	Colombia	South America	4,945,448
22	Shenyang	China	Asia	4,669,737
23	Santiago	Chile	South America	4,640,635
24	Kolkata (Calcutta)	India	Asia	4,399,819
25	St. Petersburg	Russia	Europe	4,232,105
26	Wuhan	China	Asia	4,040,113
27	Guangzhou	China	Asia	3,935,193
28	Chennai (Madras)	India	Asia	3,841,396
29	Baghdad	Iraq	Asia	3,841,268
30	Pusan	South Korea	Asia	3,814,325

Note: Population figures are for latest available years and are for defined cities. The metropolitan areas of most of these cities are much larger.
Source: 1997 Demographic Yearbook, United Nations.

MAJOR LAKES OF THE WORLD

Lake	Location	Surface Area Sq. Mi.	Surface Area Sq. Km	Maximum Depth Feet	Maximum Depth Meters
Caspian Sea	Central Asia	152,239	394,299	3,104	946
Superior	North America	31,820	82,414	1,333	406
Victoria	East Africa	26,828	69,485	270	82
Huron	North America	23,010	59,596	750	229
Michigan	North America	22,400	58,016	923	281
Aral	Central Asia	13,000	33,800	223	68
Tanganyika	East Africa	12,700	32,893	4,708	1,435
Baikal	Russia	12,162	31,500	5,712	1,741
Great Bear	North America	12,000	31,080	270	82
Nyasa	East Africa	11,600	30,044	2,316	706
Great Slave	North America	11,170	28,930	2,015	614
Chad	West Africa	9,946	25,760	23	7
Erie	North America	9,930	25,719	210	64
Winnipeg	North America	9,094	23,553	204	62
Ontario	North America	7,520	19,477	778	237
Balkhash	Central Asia	7,115	18,428	87	27
Ladoga	Russia	7,000	18,130	738	225
Onega	Russia	3,819	9,891	361	110
Titicaca	South America	3,141	8,135	1,214	370
Nicaragua	Central America	3,089	8,001	230	70
Athabaska	North America	3,058	7,920	407	124
Rudolf	Kenya, East Africa	2,473	6,405	240	73
Reindeer	North America	2,444	6,330	720	220
Eyre	South Australia	2,400	6,216	varies	varies
Issyk-Kul	Central Asia	2,394	6,200	2,297	700
Urmia	Southwest Asia	2,317	6,001	49	15
Torrens	Australia	2,200	5,698	—	—
Vänern	Sweden	2,141	5,545	322	98
Winnipegosis	North America	2,086	5,403	59	18
Mobutu Sese Seko	East Africa	2,046	5,299	180	55
Nettilling	North America	1,950	5,051	—	—

Note: The sizes of some lakes vary with the seasons.

MAJOR RIVERS OF THE WORLD

River	Region	Outflow	Approximate Length Miles	Km.
Nile	North Africa	Mediterranean Sea	4,180	6,690
Mississippi-Missouri-Red Rock	North America	Gulf of Mexico	3,710	5,970
Yangtze Kiang	East Asia	China Sea	3,602	5,797
Ob	Russia	Gulf of Ob	3,459	5,567
Yellow (Huang He)	East Asia	Gulf of Chihli	2,900	4,667
Yenisei	Russia	Arctic Ocean	2,800	4,506
Paraná	South America	Río de la Plata	2,795	4,498
Irtish	Russia	Ob River	2,758	4,438
Congo	Africa	Atlantic Ocean	2,716	4,371
Heilong (Amur)	East Asia	Tatar Strait	2,704	4,352
Lena	Russia	Arctic Ocean	2,652	4,268
Mackenzie	North America	Beaufort Sea	2,635	4,241
Niger	West Africa	Gulf of Guinea	2,600	4,184
Mekong	Asia	South China Sea	2,500	4,023
Mississippi	North America	Gulf of Mexico	2,348	3,779
Missouri	North America	Mississippi River	2,315	3,726
Volga	Russia	Caspian Sea	2,291	3,687
Madeira	South America	Amazon River	2,012	3,238
Purus	South America	Amazon River	1,993	3,207
São Francisco	South America	Atlantic Ocean	1,987	3,198
Yukon	North America	Bering Sea	1,979	3,185
St. Lawrence	North America	Gulf of St. Lawrence	1,900	3,058
Rio Grande	North America	Gulf of Mexico	1,885	3,034
Brahmaputra	Asia	Ganges River	1,800	2,897
Indus	Asia	Arabian Sea	1,800	2,897
Danube	Europe	Black Sea	1,766	2,842

River	Region	Outflow	Approximate Length	
			Miles	Km.
Euphrates	Asia	Shatt-al-Arab	1,739	2,799
Darling	Australia	Murray River	1,702	2,739
Zambezi	Africa	Mozambique Channel	1,700	2,736
Tocantins	South America	Pará River	1,677	2,699
Murray	Australia	Indian Ocean	1,609	2,589
Nelson	North America	Hudson Bay	1,600	2,575
Paraguay	South America	Paraná River	1,584	2,549
Ural	Russia	Caspian Sea	1,574	2,533
Ganges	Asia	Bay of Bengal	1,557	2,506
Amu Darya (Oxus)	Asia	Aral Sea	1,500	2,414
Japurá	South America	Amazon River	1,500	2,414
Salween	Asia	Gulf of Martaban	1,500	2,414
Arkansas	North America	Mississippi River	1,459	2,348
Colorado	North America	Gulf of California	1,450	2,333
Dnieper	Russia	Black Sea	1,419	2,284
Ohio-Allegheny	North America	Mississippi River	1,306	2,102
Irrawaddy	Asia	Bay of Bengal	1,300	2,092
Orange	Africa	Atlantic Ocean	1,300	2,092
Orinoco	South America	Atlantic Ocean	1,281	2,062
Pilcomayo	South America	Paraguay River	1,242	1,999
Xi Jiang	East Asia	China Sea	1,236	1,989
Columbia	North America	Pacific Ocean	1,232	1,983
Don	Russia	Sea of Azov	1,223	1,968
Sungari	East Asia	Amur River	1,215	1,955
Saskatchewan	North America	Lake Winnipeg	1,205	1,939
Peace	North America	Great Slave River	1,195	1,923
Tigris	Asia	Shatt-al-Arab	1,180	1,899

THE HIGHEST PEAKS IN EACH CONTINENT

Continent	Mountain	Location	Height	
			Feet	Meters
Asia	Everest	Tibet & Nepal	29,028	8,848
South America	Aconcagua	Argentina	22,834	6,960
North America	McKinley	Alaska	20,320	6,194
Africa	Kilimanjaro	Tanzania	19,340	5,895
Europe	Elbrus	Russia & Georgia	18,510	5,642
Antarctica	Vinson Massif	Ellsworth Mountains	16,066	4,897
Australia	Kosciusko	New South Wales	7,316	2,228

Note: The world's highest sixty-six mountains are all in Asia.

MAJOR DESERTS OF THE WORLD

Desert	Location	Approximate area		Type
		Sq. miles	Sq. km.	
Antarctic	Antarctica	5,400,000	14,002,200	polar
Sahara	North Africa	3,500,000	9,075,500	subtropical
Arabian	Southwest Asia	1,000,000	2,593,000	subtropical
Great Western (Gibson, Great Sandy, and Great Victoria)	Australia	520,000	1,348,360	subpical
Gobi	East Asia	500,000	1,296,500	cold winter
Patagonian	Argentina, South America	260,000	674,180	cold winter
Kalahari	Southern Africa	220,000	570,460	subtropical
Great Basin	Western United States	190,000	492,670	cold winter
Thar	South Asia	175,000	453,775	subtropical
Chihuahuan	Mexico	175,000	453,775	subtropical
Karakum	Central Asia	135,000	350,055	cold winter
Colorado Plateau	Southwestern United States	130,000	337,090	cold winter
Sonoran	United States and Mexico	120,000	311,160	subtropical
Kyzylkum	Central Asia	115,000	298,195	cold winter
Taklimakan	China	105,000	272,265	cold winter
Iranian	Iran	100,000	259,300	cold winter
Arctic	Arctic Circle	62,000	161,000	polar
Simpson	Eastern Australia	56,000	145,208	subtropical
Mojave	Western United States	54,000	140,022	subtropical
Atacama	Chile, South America	54,000	140,022	cold coastal
Namib	Southern Africa	13,000	33,709	cold coastal

Highest Waterfalls of the World

Waterfall	Location	Source	Height Feet	Height Meters
Angel	Canaima National Park, Venezuela	Rio Caroni	3,212	979
Tugela	Natal National Park, South Africa	Tugela River	3,110	948
Utigord	Norway	glacier	2,625	800
Monge	Marstein, Norway	Mongebeck	2,540	774
Mutarazi	Nyanga National Park, Zimbabwe	Mutarazi River	2,499	762
Yosemite	Yosemite National Park, California, U.S.	Yosemite Creek	2,425	739
Espelands	Hardanger Fjord, Norway	Opo River	2,307	703
Lower Mar Valley	Eikesdal, Norway	Mardals Stream	2,151	655
Tyssestrengene	Odda, Norway	Tyssa River	2,123	647
Cuquenan	Kukenan Tepuy, Venezuela	Cuquenan River	2,000	610
Sutherland	Milford Sound, New Zealand	Arthur River	1,904	580
Kjell	Gudvanger, Norway	Gudvangen Glacier	1,841	561
Takkakaw	Yoho National Park, British Columbia, Canada	Takkakaw Creek	1,650	503
Ribbon	Yosemite National Park, California, U.S.	Ribbon Stream	1,612	491
Upper Mar Valley	near Eikesdal, Norway	Mardals Stream	1,536	468
Gavarnie	near Lourdes, France	Gave de Pau	1,388	423
Vettis	Jotunheimen, Norway	Utla River	1,215	370
Hunlen	British Columbia, Canada	Hunlen River	1,198	365
Tin Mine	Kosciusko National Park, Australia	Tin Mine Creek	1,182	360
Silver Strand	Yosemite National Park, California, U.S.	Silver Strand Creek	1,170	357
Basaseachic	Baranca del Cobre, Mexico	Piedra Volada Creek	1,120	311
Spray Stream	Lauterburnnental, Switzerland	Staubbach Brook	985	300
Fachoda	Tahiti, French Polynesia	Fautaua River	985	300
King Edward VIII	Guyana	Courantyne River	850	259
Wallaman	near Ingham, Australia	Wallaman Creek	844	257
Gersoppa	Western Ghats, India	Sharavati River	828	253
Kaieteur	Guyana	Rio Potaro	822	251
Montezuma	near Rosebery, Tasmania	Montezuma River	800	240
Wollomombi	near Armidale, Australia	Wollomombi River	722	2203

Source: Fifth Continent Australia Pty Limited.

INDEX

Endogenic sediment, 379

Energy, 379; alternative sources, 266-270, 323; and pollution, 265; sources, 257-265; tidal power, 269; and warfare, 263; wind power, 268

Engineering projects, 270-275; environmental problems, 273

English Channel; and Chunnel, 274

Environment, 379

Environmental degradation, 379

Environmental determinism, 6, 379

Environmental ethics, 380

Environmental justice, 380

Environmental Literacy Council, 144

Eocene epoch, 380

Eolian, 380

Eolian deposits, 380

Eolian erosion, 380

Eon, 380

Epeiric sea, 380

Epicenter, 380

Epicontinental sea, 381

Epifauna, 381

Epilimnion, 381

Epoch, 381

Equal-area projection, 381

Equator, 381; and climate, 110, 135, 149, 198; and seasons, 41, 106

Equinox, 381

Equinoxes, 39, 41

Era, 381

Eratosthenes, 3, 10

Erg, 381

Erie Canal, 272

Eritrea, 313

Erosion, 73, 79, 93, 161, 381; and agriculture, 246-247, 257; and deforestation, 265; eolian, 91; and glaciation, 84; and mountains, 71; and ocean patterns, 95, 97-98; and overgrazing, 166, 181, 247; and sinkholes, 82. *See also* Fluvial processes

Eruption, volcanic, 381

Escarpment, 382

Esker, 382

Estuarine zone, 382

Estuary, 382

Etesian winds, 382

Ethnic group, 382

Ethnic religion, 382

Ethnocentrism, 382

Ethnography, 382

Europe; air pollution, 209; canals, 273; minerals, 279; national parks, 159; railroads, 228-229; roads, 224

Eustacy, 382

Eustatic movement, 382

Evaporation, 382

Evapotranspiration, 382

Exclave, 382

Exfoliation, 383

Exosphere, 383

Expansion-contraction cycles, 383

Exploration; and early geographers, 5; world, 214-220

External economies, 383

Extinction, 384

Extrusive rock, 384

Exxon Valdez, 280

Fahrenheit scale, 384

Fall line, 384

Famine, 384

Fata morgana, 384

Fathom, 385

Fathometer, 385

Faults, 355, 361, 372, 376, 380, 385-386, 392, 421, 500, 523, 538; and grabens, 399; and horst, 404; normal, 455; reverse, 487; and rifts, 488; and ring dikes, 489; scarp, 382, 497; slip-strike, 516; thrust, 527; transform, 532

Fauna, 386

Feldspar, 324, 386, 399

Fell, 386

Felsic rocks, 386

Fen, 386

Feng shui, 386

Fenno-Scandian Shield, 87

Fertility rate, 386

Fertilizer, 386

Fetch, 387

Population pyramid, 475
Porosity, 475
Ports, 475; Africa, 229; North America, 294
Possibilism, 6, 475
Postindustrial economy, 475
Potable water, 475
Potato famime, Irish, 199
Potatoes, South America, 253
Potholes, 475
Powell, John Wesley, 332
Powell, Lake, 429
Prairie, 476
Prairie dogs, 147
Precambrian period, 476
Precipitation, 477; in desert climates, 110-111, 135, 148-149, 151; in grasslands, 146-147; and monsoons, 110; and topography, 198
Primary economic activity, 477
Primary wave, 477
Primate cities, 477, 484
Prime meridian, 12, 195-197, 477; Ptolemy's, 10
Principal parallels, 478
Principality, 478
Protectorate, 478
Proterozoic eon, 478
Protruded, 478
Province, 478
Psychrometer, 478
Ptolemy, 4-5, 10
Public Land Survey, U.S., 13
Pull factors, 478
Pumice, 57, 478
Push factors, 479
Pyroclasts, 479

Qanat, 479
Quartz, 480
Quaternary period, 480

Radar imaging, 480
Radial drainage, 480
Radiation, 480
Radio, 105
Radioactivity, 456, 480-481; and carbon dating, 341

Radon gas, 481
Railroads, 225-230; and time zones, 195-197; and urbanization, 192
Rain forests, 170, 481; medicinal plants, 171; tropical, 140-141, 482
Rain gauge, 482
Rain-shadow effect, 482
Rainfall, 482
Rainier, Mount, 514
Rank-size rule, 484
Rapids, 484
Ratzel, Friedrich, 308, 324
Realm, 484
Recessional moraine, 484
Recumbent fold, 484
Red Sea, 52, 72, 422
Reef (geology), 485
Reefs, 329, 485; coral, 357; fringing, 392
Refraction of waves, 485
Region, 485
Regionalism, 485
Regolith, 485
Regression, 485
Rejuvenation, 485
Relief, 486
Religion, 486
Remote sensing, 486
Renewable resources, 486
Replacement rate, 486
Republic, 486
Reservoir, 486
Reservoir rock, 486
Resources, 486; nonrenewable, 174-175, 177-178, 455; renewable, 170-174; strategic, 513
Respiration, 487
Retrograde orbit, 487
Retrograde rotation, 487
Reverse fault, 487
Revolution, 488
Ria coast, 488
Rice, 243, 253; Asia, 219, 253; North America, 199-200, 243
Richter, Charles, 488
Richter scale, 488
Ridge, 488
Rift, 488
Rift propagation, 488

Tropopause, 104, 534
Troposphere, 104, 534
Trough, 534
Tsunamis, 182, 534-535
Tufa, 535
Tuff, 535
Tuff ring, 535
Tumescence, 535
Tundra, 134-135, 153-156, 535;
    conservation of, 155; fauna, 154
Tunnel vent, 536
Tunnels, 225, 273-274; natural, 89;
    and railroads, 227
Turbulence, 536
Turtles, sea, 173
Typhoons, 122, 536

Ubac slope, 537
Ultimate base level, 537
Ultramafic rocks, 537
Ultraviolet radiation, 101-102, 104,
    537; and ozone, 208
Uluru, 415
Unconformity, 537
Underclass, 538
Underemployment, 538
Uniform Time Act, 196
Uniformitarianism, 65, 538
United Arab Emirates, 313
United Nations, and national parks,
    157
United States; borders, 312; and
    China, 310; climate and
    productivity, 200; commercial
    agriculture, 242-244; government
    of, 299-300; iron ore, 278; labor
    force, 282-283, 286; minerals,
    278; national parks, 156;
    railroads, 228-229; roads, 221,
    223; water resources, 167
United States Naval Observatory,
    198
Universal Transverse Mercator, 538
Universalizing religion, 538
Upland, 538
Uplift, 538
Upper mantle, 538
Upthrown block, 538
Upwelling, 538

Uranus, 29
Urban heat islands, 194, 539
Urbanization, 539; and agriculture,
    190, 249; counterurbanization,
    359; and food distribution, 254;
    global, 190-194;
    overurbanization, 460; and
    population growth, 189;
    suburbanization, 517
Utah, 391
UTC. *See* Coordinated universal time
Uvala, 539

Vadose zone, 539
Valley train, 540
Valleys, 364, 373, 394, 539; drowned,
    375; fjords, 387; rift, 488; river,
    490; transverse, 533; u-shaped,
    536
Van Allen radiation belts, 540
Varnish, desert, 540
Varve, 540
Vatican City, 447
Vegetation, 540
Vein, 540
Veld, 540
Venezuela, roads, 223
Ventifacts, 540
Venus, 27
Vernacular region, 540
Vernal equinox, 540
Verrazano-Narrows Bridge, 272
Victoria, Lake, 312
Victoria Desert, 149
Vietnam, and China, 313
Vikings, and ice age, 88
Village, 540
Volcanoes, 45, 50, 53, 55-63, 70-71,
    541; ash, 540; and atmosphere,
    101, 207; and earthquakes, 541;
    Hawaii, 54; Iceland, 55; and
    island arcs, 541; on other planets,
    26; South America, 70
Volcanology, 542
Vostok, 113

Wadi, 542
Warfare, and energy, 263
Wash, 542